Covering Extended Reality Technologies in the Media

This book presents a study of the news coverage of extended reality technologies (virtual, augmented and mixed reality; or XR) and how this news corresponds with the marketing of XR products.

Focusing on a group of recently emerging technological products, the book offers in-depth analysis of the news coverage of XR technologies and explores the overlap between news discourse and promotional discourse by comparing the way these products are framed in the news and their marketing materials. Using both quantitative and qualitative data, it discusses the topics covered in XR news, as well as the sources used and the specific framing techniques that appear in both XR news and marketing materials. In addition to these findings, it also provides a set of frame categories that can be used by other researchers analysing the media coverage of emerging technologies.

Ultimately arguing that the news represents XR in such a way that treats readers as consumers instead of citizens, prioritising the interests of XR companies rather than news audiences, this book will be of interest to students and researchers in media and communications, discourse studies, journalism, PR and marketing and innovation studies, as well as XR practitioners.

Emma Kaylee Graves is Senior Lecturer in Media and Communications at Canterbury Christ Church University.

Routledge Research in Journalism

For more information about this series, please visit: www.routledge.com/Routledge-Research-in-Journalism/book-series/RRJ

Covering Extended Reality Technologies in the Media

Emma Kaylee Graves

Routledge
Taylor & Francis Group

LONDON AND NEW YORK

First published 2024
by Routledge
4 Park Square, Milton Park, Abingdon, Oxon OX14 4RN

and by Routledge
605 Third Avenue, New York, NY 10158

Routledge is an imprint of the Taylor & Francis Group, an informa business

© 2024 Emma Kaylee Graves

British Library Cataloguing-in-Publication Data
A catalogue record for this book is available from the British Library

Library of Congress Cataloging-in-Publication Data
Names: Graves, Emma Kaylee, author.
Title: Covering extended reality technologies in the media / by Emma Kaylee Graves.
Description: London ; New York : Routledge, 2024. |
Series: Routledge research in journalism |
Includes bibliographical references and index.
Identifiers: LCCN 2023038449 (print) | LCCN 2023038450 (ebook) |
ISBN 9781032446646 (hardback) | ISBN 9781032451886 (paperback) |
ISBN 9781003375814 (ebook)
Subjects: LCSH: Mixed reality–Press coverage. |
Human-computer interaction–Press coverage. |
Technological innovations–Press coverage.
Classification: LCC QA76.9.H85 G719525 2024 (print) |
LCC QA76.9.H85 (ebook) | DDC 006–dc23/eng/20231004
LC record available at https://lccn.loc.gov/2023038449
LC ebook record available at https://lccn.loc.gov/2023038450

ISBN: 978-1-032-44664-6 (hbk)
ISBN: 978-1-032-45188-6 (pbk)
ISBN: 978-1-003-37581-4 (ebk)

DOI: 10.4324/9781003375814

Typeset in Sabon
by Newgen Publishing UK

Contents

Illustrations

Foreword

In what ways, we might ask, do inflections of extended reality (XR) – a term increasingly being used to describe the varied, uneven interweaving of virtual reality (VR), augmented reality and mixed reality technologies – influence our perceptions of reality? Much depends on what we mean by "reality", of course, rather to the annoyance of those who refuse to countenance any blurring of what they would insist are clear empirical divisions between our vision and the world around us.

In reading Emma Kaylee Graves's richly insightful *Covering Extended Reality Technologies in the Media*, I found myself thinking about the etymology of VR and its entanglements over the years. Perhaps not surprisingly where emergent technologies are concerned, fictional depictions came to mind in the first instance, such as Ray Bradbury's 1953 novel *Fahrenheit 451*, or William Gibson's *Neuromancer* and Pamela Sargent's *The Shore of Women* in the mid-1980s, amongst numerous other, typically dystopian imaginings. From a scholarly vantage point, just as the term was gaining popular purchase in media commentaries, Jonathan Crary published his book *Techniques of the Observer* (MIT Press, 1990). This intervention quickly proved formative for efforts to theorise the lineages of VR's visualities. Crary usefully sought to historicise what he characterised as a transformation in the nature of visuality across centuries – that is, the gradual, still inchoate consolidation of new forms of visual representation and perception – becoming increasingly apparent as analogue realms gave way to digital ones. In describing the "sweeping reconfiguration of relations between an observing subject and modes of representation" underway, Crary queried the extent to which older, more familiar modes of "seeing" would persist, or "coexist uneasily" alongside those being introduced in the fabricated visual "spaces" associated with computer-generated imagery (1990: 1–2).[1]

The emphasis Crary placed on considering "problems of vision" as "fundamentally questions about the body and the operation of social power" inspired fresh thinking about the status of the individual reliant upon seeing technologies as an observer in their own right. "Vision and its effects are always inseparable from the possibilities of an observing subject", he argued, "who is both the historical product and the site of certain practices, techniques, institutions, and procedures of subjectification" (1990: 5). In recognising how the observer "sees within a prescribed set of possibilities", and is thereby "embedded in a system of conventions of limitations", certain seemingly mundane, everyday "rules, codes, regulations and practices" came to the fore (1990: 6). The observer, it follows, is necessarily enmeshed within this shifting, evolving field of perception. Any claim to their positionality outside or external to the "hegemonic organization of the visual" is thus rendered problematic, not least where "sensory discipline" is concerned. Whether it is the camera obscura, stereoscope, kaleidoscope, phenakistiscope or diorama being considered, attention to the historical specificities of vision prompts questions of embodiment and subjectivity. Accordingly, Crary's call to move beyond technological determinism to recognise the lived dynamics of social power continues to resonate today, much to the advantage of creative attempts to reenvisage in more open, progressive ways the promise of virtual worlds for everyone.

Making good this promise is a formidable challenge precisely because of the types of interrelated factors Graves brings to light in *Covering Extended Reality Technologies in the Media*. In choosing to delve into the framing strategies giving shape and direction to pertinent news reporting of XR, she reveals how marketing discourses of consumerism so often prevail on the pages of the British newspapers under scrutiny, inviting concerns about the quality – indeed, integrity – of press performance. The elucidation of key drivers helps to disrupt certain familiar assumptions about the media's influence on technological acceptance and diffusion, especially where the ideological mediation of promotional framings of innovation and use-value proves salient. News coverage will be ethically compromised, Graves's findings show us, to the extent the commercial interests of XR companies are prioritised over and above the public's need for informed, rigorous reporting of potential risks and harms as well as possible benefits. This study's conclusions advance our understanding of the issues at stake in important ways, while also illuminating conceptual and methodological resources to facilitate future enquiries into the ongoing commercialisation of technology news – and what we can do to improve journalistic standards in the public interest.

Stuart Allan
School of Journalism, Media and Culture
Cardiff University, UK

Note

1 Crary, J. (1990) *Techniques of the Observer*. Cambridge, Massachusetts: MIT Press.

Acknowledgements

For her support and encouragement from the beginning of this project, I extend my sincere thanks to Dr Ruth Sanz Sabido. Thanks also go to Dr Andrew Butler and Professor Agnes Gulyas for their input during various important stages. I would like to thank Professor Stuart Allan and Dr Alan Meades for their valuable insights that allowed me to enhance this work. Further gratitude goes to Stuart Allan in particular for writing the foreword of this book. Finally, for their feedback and advice on this monograph, I thank Professor Angela Pickard and Professor Chris Pallant.

1 Introduction

Much of what we know about new technologies and innovations comes from the media, particularly when commercial products are in their pre-release stages (Chuan, Tsai and Cho, 2019; Cogan, 2005; Kelly, 2009; McKernan, 2013; Scheufele and Lewenstein, 2005; Sun et al., 2020). The companies creating these products aim to convince the general public that they are worth buying, communicating with their audiences using marketing and public relations strategies. Whereas marketing and public relations ultimately have a profit-seeking agenda, the news media are supposed to hold the interests of the public above all others by providing critical coverage that allows readers to understand both the benefits and potential drawbacks or risks regarding new products (Anderson et al., 2005; Hansen, 2018; Kovach and Rosenstiel, 2014; McNair, 2009; Schäfer, 2017; Scheufele, 2013). However, factors such as news production practices and commercialisation mean that this does not always happen.

Along those lines, on 6 March 2015, the online edition of *T3* magazine posted an article with the headline "Better than life: 2015's hottest VR, console and PC gaming tech" (*T3 Online*, 2015). The article introduced some upcoming virtual reality (VR) products, including HTC Vive, PlayStation VR (then named Project Morpheus) and motion capture peripherals that could be used with VR headsets. It was troubling to me that this article was encouraging escapism into immersive virtual worlds by insinuating the experience would be better than real life. Jaron Lanier, who is credited with coining the term "virtual reality" (Rheingold, 1991), envisioned that the technology would improve upon the real world, rather than offer a compelling alternative. He explains: "When my friends and I built the first virtual reality machines, the whole point was to make this world more creative, expressive, empathic, and interesting. It was not to escape it" (Lanier, 2011: 33). The "better than life" phrase used in the *T3* article contested Lanier's original vision, instead risking disillusionment with the real world. Therefore, this article was the initial inspiration for researching news coverage of VR to uncover whether this was a one-off case or if such sentiments were more

DOI: 10.4324/9781003375814-1

widespread, particularly in traditional news coverage that is supposed to provide critical and impartial information to citizens.

My earlier studies (carried out before the work presented in this book) revealed that, while these news articles rarely referred to VR as superior to real life, news coverage of VR was largely very positive and some articles even prompted readers to purchase these products alongside links to relevant retailers (Graves, 2016, 2017). These findings raised questions about the extent to which the news acts as a promotional tool for new technologies (such as VR) rather than maintaining a clear boundary between news and advertising content. This led to an extensive study on that topic, culminating in the research discussed here.

This book presents the first in-depth investigation into the news of extended reality (XR) technologies (encompassing VR, augmented reality (AR) and mixed reality (MR)) and how it relates to the marketing of these products. The study utilises framing theory from a social constructivist perspective as its main theoretical approach, which informs a rigorous mixed methods research design combining content analysis and framing analysis. Using these quantitative and qualitative approaches, the study investigates three main areas: (1) how the news frames XR; (2) the extent to which these frames align with those in XR promotional materials; and (3) whether the way the news frames XR could promote its diffusion. This introductory chapter sets out how the book contributes to the academic literature, as well as its implications outside of academia for XR professionals. It then defines the key concepts that are related to the research, justifies the topic of focus and provides some brief methodological information, before outlining the content of the rest of the book.

Areas of Contribution

This book makes three key original contributions to the field. First, it contributes to the existing academic literature on news coverage of emerging technologies by analysing a topic that had previously been unexplored – XR. Although one published study had examined news coverage of the Pokémon Go AR game (Grandinetti and Ecenbarger, 2018), the authors provided very little detail regarding their methodology and selection of news reports. In addition to being limited to just one AR application, this raises questions as to the reliability of their findings. This book presents the first study that has looked at XR news with a wide scope.

Second, it contributes to knowledge about the relationship between news and promotional content. Previously, the research on this topic has only focused on two areas: the use of native advertising and the reliance on press releases (e.g. Chyi and Lee, 2018; Erjavec, 2004; Harro-Loit and Saks, 2006; Lewis, Williams and Franklin, 2008; Pander Maat, 2007; Sissons, 2012). Significantly, this book considers a broader and more extensive sample of marketing materials (e.g. press releases plus websites, social media and video advertisements) and their relation to news content, providing deeper insights in this area.

Third, aside from these empirical contributions, the book makes both a theoretical and methodological contribution by developing a set of frames and frame categories for researching the media coverage of emerging technologies. The specific frames that have been identified can be used in other studies focusing on XR and, extending beyond this, the frame categories can be used as guidance in future research on the media coverage of different emerging technologies. That is to say, scholars can use this categorisation to identify the frames that are applied to other emerging technologies. For studies of innovations, it is important to be able to identify frames inductively rather than use a prescriptive set of frames since new technologies may bring with them new concepts. However, such approaches have been criticised for lacking replicability and comparability (e.g. Tankard, 2001). The development of these frame categories means that researchers can maintain the benefits of identifying frames unique to their context while avoiding these criticisms, since individual frames will come under categories that can be compared across studies.

Furthermore, both the data and model presented in this book can be of use to practitioners outside of academia. The data can help industry practitioners understand how XR technology has been portrayed to their audiences, which could inform professional practices. Additionally, the model of frame categories could be used by professionals to carry out market research to better understand their audiences, not just for XR but for other technologies as well.

Defining "News"

Before going further, it is important to define some of the key concepts in this book, starting with what is meant by "news". The definition of the concept of "news" has been much-debated and theorised, both in relation to what news perhaps *should* be and how journalists decide whether a story is newsworthy enough to become published, thus making it a "news" item. For instance, Bednarek and Caple note that the journalism profession is "built on the values of objectivity, fairness, truthfulness and accuracy" (2012: 36). Indeed, in the UK context that this book focuses on, the National Union of Journalists' Code of Conduct states that a journalist should produce news that is honest, accurate and differentiates between fact and opinion (National Union of Journalists, 2018). Alternatively, news might be defined in relation to how stories are selected by journalists. The concept of news values is often referred to here, which was first theorised by Galtung and Ruge (1965) and subsequently developed by several authors (including Bell, 1991; van Dijk, 1988; Gans, 1980; Harcup and O'Neill, 2001, 2017). While news values will be discussed in more detail in Chapter 3, broadly, this theory suggests that journalists determine whether a story is newsworthy based on certain factors, including significance, novelty, reference to elites, human interest, negativity and so on. This could then shed some light on what "news" is.

This book does not attempt to theorise a definition of news; however, it is important to specify how the concept of news was treated in the research. For the purpose of sample collection, "news" was defined as any articles published by the analysed news outlets, with the exception of review articles. This was based on the idea that news is supposed to be objective or impartial (Bednarek and Caple, 2012; Sambrook, 2012), whereas reviews are clearly presented to the audience as including opinions. In other words, the audience would be less likely to expect to see facts or objective information in a review than they would in a text that is presented to them as a "news" article on a news site. While it is acknowledged that not all news is objective (indeed, the results of the current study certainly evidence this, and others argue that objectivity is impossible (e.g. Davies, 2009; Maras, 2013; Muñoz-Torres, 2012)), the aim was to analyse content that was presented to the audience as "news", thus being associated with the ideas of independence and truth-telling that might be expected from news outlets. For the purpose of analysis, the concepts of news values and best practices in journalism will be used to discuss the news articles.

Defining "Extended Reality Technologies"

As noted above, this book deals with a group of technologies that can be categorised as "extended reality technologies" or XR. This is a fairly broad term that encompasses a range of technologies that "extend" the user's perception of reality by allowing them to view virtual worlds and/or objects. While these technologies are sometimes termed "immersive technologies", this book deliberately avoids referring to them in this way because it argues that these products are not inherently immersive (as will be discussed in Chapter 4). In particular, this book focuses on virtual, augmented and mixed reality technologies, which can be defined as follows.

While the technologies under the XR umbrella differ in the way they treat the physical environment, they are similar in that they each modify (or *extend*) reality in some form. In 1994, Milgram and Kishino proposed the concept of a "virtuality continuum" to classify different types of VR. The authors define the VR environment as "a completely synthetic world" that the user is "totally immersed in, and able to interact with" (1994: 2). Indeed, immersion is a key aspect of the technologies, which will be discussed in Chapter 2. Milgram and Kishino see AR and MR as subsets of VR. They define AR as "any case in which an otherwise real environment is 'augmented' by means of virtual (computer graphic) objects" (1994: 4). Furthermore, MR is defined as the combination of the real and virtual environments.

These definitions have mostly remained consistent in recent years. Indeed, Brigham states that VR "obscures the user's physical surroundings and replaces them with a computer-generated scene or one that was previously captured" (2017: 173). On the other hand, AR "allows a person to see the real, physical world, but it is overlaid with a layer of digital content in real time" (Brigham, 2017: 172). Similarly, MR "allows a person to see the real, physical world and objects but also see believable, and even responsive, virtual objects" (Brigham, 2017: 174). There is clear overlap between AR and MR and these terms are sometimes used interchangeably (Carter and Egliston, 2020). However, the main difference between the two is that the digital objects seen with MR are able to *interact* with the physical environment, whereas the digital elements displayed using AR are simply superimposed on top of the physical environment (Brigham, 2017; Greengard, 2019). In other words, MR is somewhat more advanced than AR in the way that it treats the physical environment and user interaction.

In 2023, virtual, augmented and mixed reality can currently be experienced in two main ways: through the use of a smartphone or wearing a head-mounted display (HMD). In the years covered by this study (2012–2017), MR was only accessible through the use of an HMD. Such devices include Microsoft HoloLens and Magic Leap. On the other hand, AR can be experienced either by using an HMD (such as Google Glass) or through smartphone applications that utilise the device's camera (Greengard, 2019). Additionally, the VR devices of focus in this study always involve the use of an HMD. However, there are some variations in how this can work. Users may wear a dedicated headset that includes all the technology needed for the experience in the device itself, or they may use a cheaper headset that utilises a smartphone as the screen (Evans, 2019). During the sample period of this study, dedicated headsets required connection to an external power source to function. These devices include the Meta-owned (or, as it was known at the time, Facebook-owned) Oculus Rift, Sony's PlayStation VR and the HTC Vive. However, since 2018, standalone headsets have been released that are comparable in quality to the previous dedicated headsets but with the benefit of functioning without being connected to an external power source. The first such device was Oculus Go, released in May 2018 (Evans, 2019). This demonstrates that XR can take several forms and is still evolving after its initial inception. The next chapter will provide some more detail about XR, including the state of the industry and its applications.

Why Media Discourse?

As mentioned above, this study utilises framing theory from a social constructivist perspective. Both of these theories highlight the power that language (including news discourse) can have in constructing reality. For instance, regarding framing, Pan and Kosicki state that choices of words "hold great power in setting the context for debate, defining issues under consideration, summoning a variety of mental representations, and providing the basic tools to discuss the issues at hand" (1993: 70). That is to say, what language is used, and how it is used, works towards framing – and thus constructing – reality in certain ways. This aligns with social constructivism which sees reality as created through interaction with others in a social system, rather than existing in an objective form (Slater, 2017). Indeed, Hallahan argues that "[f]raming is a critical activity in the construction of social reality because it helps shape the perspectives through which people see the world" (1999: 207). Along the same lines as Pan and Kosicki, Richardson (2007) also stresses the power of news discourse in contributing to the construction of reality. The author states that journalism "can help shape social reality by shaping our views of social reality. For these reasons, and many more, the language of the news media needs to be taken very seriously" (Richardson, 2007: 13). Therefore, analysing the framing of XR can reveal how the technology has been socially constructed.

Regarding emerging technologies specifically, McKernan argues that "nascent technologies provide opportunities for different discursive outlets to construct or reiterate powerful cultural codes and worldviews" (2013: 309). Certainly, when it comes to new technologies, the news media are particularly powerful in shaping public attitudes and opinions, since most individuals have little or no knowledge about these innovations (Chuan, Tsai and Cho, 2019; Cogan, 2005; Dimopoulos and Koulaidis, 2002; Hetland, 2012; Kelly, 2009; Royal, 2006; Scheufele and Lewenstein, 2005). This is especially the case since the mass media are the public's main source of information about emerging technologies (Cacciatore et al., 2012; Sun et al., 2020; Whitton and Maclure, 2015; Williams, 2003). Indeed, Scheufele and Lewenstein (2005) found that respondents who were frequent readers of nanotechnology news (which is mostly positive) were more likely to believe the benefits of the technology outweighed the risks than those who were not frequent readers of this news. Moreover, Buenaflor and Kim argue that the "perception of a new technology significantly affects acceptance" (2013: 107). Therefore, as the news media are a major force affecting public perception of emerging technologies, how they represent these products could ultimately impact their adoption.

In other areas, previous research has uncovered a blurring of the boundary between news and promotional content (Chyi and Lee, 2018; Erjavec, 2004; Harro-Loit and Saks, 2006; Lewis, Williams and Franklin, 2008; Pander Maat, 2007; Sissons, 2012). Such practices "compromis[e] the independence of the press" (Lewis, Williams and Franklin, 2008: 2). While "product promotion aims at manufacturing a favorable view toward a product" (Chyi and Lee, 2018: 588), the purpose of news content should be to inform and educate the general public (Kovach and Rosenstiel, 2014). Thus, if similar frames appear in the news and marketing of XR products, this would indicate not only a blurring between news and advertising, but that the news has acted as a promotional tool for these technologies. This emphasises the importance of analysing the relationship between news and marketing discourse.

Why Extended Reality?

Of course, there are many different emerging technologies, including the more science-focused nanotechnology, biotechnology and genetically modified products, as well as computer-based innovations such as smart energy metres, artificial intelligence and autonomous vehicles. However, XR technologies stand out for a number of reasons, making it particularly worthwhile to examine how they have been framed in the news. First, XR has been described not just as a new technology, but as a new *medium* (Evans, 2019; Li et al., 2020; Papagiannis, 2014), bringing with it new concepts and experiences. Thus, as XR is notably different from previous technologies, analysing its news framing makes a valuable contribution to the literature.

Second, XR alters how individuals perceive reality either by immersing the user into a completely virtual environment or by overlaying digital objects on the physical environment. Because of this, representations of XR can impact the public's view towards not only the virtual but also the real (Chan, 2014). This makes its news representations even more important since their effect can extend beyond the technology itself to the wider world.

Third, regarding VR specifically, Madary and Metzinger argue the following:

> VR technology will eventually change not only our general image of humanity but also our understanding of deeply entrenched notions, such as "conscious experience," "selfhood," "authenticity," or "realness." In addition, it will transform the structure of our life-world, bringing about entirely novel forms of everyday social interactions and changing the very relationship we have to our own minds.
>
> (2016: 1–2)

If XR can have such profound effects, how the news frames this technology is vital because it could impact how many people adopt the technology and thus become susceptible to these effects. Finally, several major companies are involved in the XR industry, with Meta (or Facebook), Google, Microsoft, Samsung, HTC and more each having their own XR products. Therefore, by analysing XR news and marketing, the book is able to provide a critical analysis of how these elite organisations may impact the news.

Analysing Media Coverage of Extended Reality

The study presented in this book has two main aims: (1) to analyse the way XR has been presented in the news; and (2) to investigate the extent to which the news has acted as a promotional tool for XR. To address these points, this study is primarily underpinned by framing theory. Framing can be understood as presenting an issue or topic in a way that emphasises certain aspects, while obscures others, in the interests of promoting a particular interpretation of that issue or topic (Allan, Anderson and Petersen, 2010; de Vreese, 2010; Entman, 1993; Gitlin, 1980; Hallahan, 1999; Scheufele and Scheufele, 2010). Regarding the specific aims, the former is addressed by analysing a sample of news articles about XR to identify patterns in the coverage, as well as the key frames that are applied to the technology. The latter aim is achieved in two ways. Firstly, the marketing materials of XR products are compared to the news articles by examining the framing techniques used in both discourses. Secondly, the study analyses whether the way the news presents XR is positively or negatively related to the perceived characteristics of technological innovations that make them more likely to be adopted (Buenaflor and Kim, 2013; Davis, 1989; Kim, Chan and Gupta, 2007; Rogers, 2003).

To achieve this, a mixed methods approach was developed that combined quantitative content analysis and qualitative framing analysis. It utilised two datasets: (1) a sample of 977 online news articles about XR from the *Sun*, the *Guardian* and *MailOnline* (allowing for a comparison between traditionally tabloid, mid-market and quality outlets); and (2) a range of marketing materials (e.g. video adverts, press releases, historical versions of websites and social media posts) of some of the key XR products (Oculus Rift, Samsung Gear VR, Google Glass, Microsoft HoloLens and Magic Leap). This focused on the time period of 1 January 2012 to 31 December 2017, covering the initial announcement and release of the first VR, AR and MR products in the recent XR resurgence.

Underpinning Theoretical Models

It is also important to provide some detail about the theoretical models that underpin the discussion throughout this book. Mainly, this focuses on framing devices, the hierarchy of influences model (Shoemaker and Reese, 2014) and theories of technological diffusion. Within framing theory, it is argued that there are two stages to the framing process: frame-building and frame-setting. This book focuses on frame-building, which "occurs when journalists construct news stories out of the bits and pieces of everyday life" (Moy, Tewksbury and Rinke, 2016: 8). In other words, by choosing what to include in a news story and how to write about a topic, the journalist constructs certain frames. Entman posits that frames "are manifested by the presence or absence of certain keywords, stock phrases, stereotyped images, sources of information, and sentences that provide thematically reinforcing clusters of facts or judgements" (1993: 52). These elements that make up a frame have been termed "framing devices" (Gamson and Lasch, 1983; Linström and Marais, 2012; Pan and Kosicki, 1993). Several researchers have developed lists of framing devices (also known as signifying elements) which "are tools for newsmakers to use in composing or constructing news discourse as well as psychological stimuli for audiences to process" (Pan and Kosicki, 1993: 59). Linström and Marais (2012) produced a list that syntheses the framing devices highlighted by other authors, classified into two categories: rhetorical devices and technical devices. Put simply, rhetorical devices refer to issues of language (such as word choice, metaphors and keywords), whereas technical devices refer to the elements of a news article (such as headlines, sources and how sources are identified). Throughout this book, Linström and Marais' clear categorisation is used to analyse how news frames have been constructed by identifying the particular rhetorical and technical framing devices that appear.

Furthermore, this study is also concerned with the factors that have impacted the frame-building process. Shoemaker and Reese's (2014) hierarchy of influences model defines five levels of influence on the frame-building process: (1) social systems; (2) social institutions; (3) media organisations; (4) routine practices; and (5) individuals. However, since a textual study of news (as opposed to an ethnographic study in a newsroom) uncovers little detail about specific journalists aside from their name, position and perhaps gender, the fifth factor (individuals) is not considered here. To provide further clarification on the other factors, the first level, social systems, refers to the widest impact factor comprising of ideological and sociocultural influences. The second level (social institutions) refers to influences based on the place of a media organisation in relation to other organisations and institutions. At the third level, media organisations may influence news content based on their own ideologies, such as political leaning. Fourth, the routine practices level refers to how the newsroom routine may influence news. In this book, the hierarchy of influences model acts as a useful analytical tool to examine the factors that might impact the news framing of XR.

Lastly, as this book focuses on a study of news coverage about an emerging technology with a consideration of how it could impact adoption, theories related to the diffusion of innovations are considered. Rogers defines diffusion as "the process in which an innovation is communicated through certain channels over time among the members of a social system" (2003: 5). Of most interest for this research is the innovation-decision process which splits the potential adoption of an innovation into five stages: knowledge, persuasion, decision, implementation and confirmation (Rogers, 2003). The first two stages in this model are key in the context of news framing due to their focus on the construction of meaning. According to Rogers, during the persuasion stage, the perceived attributes of innovations are particularly important. He states that there are five characteristics of innovations that are the most important in explaining the rate of their adoption (relative advantage; compatibility; complexity; trialability; and observability). While it is not the purpose of this study to measure adoption or diffusion of XR, these five characteristics defined by Rogers are applied as a useful analytical tool to consider the way the news coverage of XR may promote or hinder the diffusion of the technology. Additionally, other models of technological acceptance are also applied in the same way, including the technology acceptance model (TAM; Davis, 1989), the value-based adoption model (VAM; Kim, Chan and Gupta, 2007) and, of particular relevance, Buenaflor and Kim's (2013) factors that impact the acceptance of wearable computers. This allows the analysis to consider the extent to which the news coverage emphasises factors of a technology that make it more likely to be adopted.

Book Structure

While this chapter has introduced the aims of the book and the research presented within it, it is important to understand some background information about the technologies under the XR umbrella before analysing its media coverage. Chapter 2 provides just that, by discussing virtual, augmented and mixed reality and exploring the histories of their development. It provides contextual information regarding the state of the XR market during the period of this study (including revenue, uses and stakeholders), while also noting more recent developments. It discusses the key concepts related to XR (immersion and presence) and explores the various benefits and ethical concerns surrounding the technologies. This is complemented by an analysis of moral panic literature, particularly focusing on technology.

Following this, Chapters 3–7 present the findings of the study. Chapter 3 primarily deals with quantitative data uncovered through the application of a coding sheet. It examines the overall characteristics of XR news coverage and how this affects the framing of XR. As noted above, this book also introduces a model of frame categories that can be applied in further research of emerging technologies. Chapters 4–7 are each based on a different category within this model, first setting out some supporting information about the category before discussing the findings related to it for the study of XR media coverage. Chapter 4 examines frames that conceptualise XR (Immersive and Transcendent). Additionally, Chapter 5 focuses on frames relating to the newness of XR (Different and Unique; Revolutionary and Transformative; and Advanced and High-Quality). Chapter 6 discusses the frames relating to the user experience of XR (Social; Easy to Use; and Comfortable). The final data analysis chapter (Chapter 7) is slightly different in its structure. It first explores the specific evaluative frames that were applied to XR in the news articles (Important; Successful; Affordable; and Much-Anticipated), followed by a section that examines the overall tone of the articles (regardless of specific frames).

Ultimately, this book argues that XR news prioritises the interests of the companies that create these technologies over the interests of the general public, neglecting journalistic principles. Chapter 8 elucidates this argument based on the research findings. The chapter then provides more detail about the frames and frame categories identified in this research that can be used as methodological guidance in future studies.

References

Allan, S., Anderson, A. and Petersen, A. (2010) 'Framing Risk: Nanotechnologies in the News', *Journal of Risk Research*, 13(1), pp. 29–44. doi:10.1080/13669870903135847.

Anderson, A., Allan, S., Petersen, A. and Wilkinson, C. (2005) 'The Framing of Nanotechnologies in the British Newspaper Press', *Science Communication*, 27(2), pp. 200–220. doi:10.1177/1075547005281472.

Bednarek, M. and Caple, H. (2012) *News Discourse*. London: Continuum International Publishing Group.

Bell, A. (1991) *The Language of News Media*. Oxford: Blackwell.

Brigham, T.J. (2017) 'Reality Check: Basics of Augmented, Virtual, and Mixed Reality', *Medical Reference Services Quarterly*, 36(2), pp. 171–178. doi:10.1080/02763869.2017.1293987.

Buenaflor, C. and Kim, H. (2013) 'Six Human Factors to Acceptability of Wearable Computers', *International Journal of Multimedia and Ubiquitous Engineering*, 8(3), pp. 103–114.

Cacciatore, M.A., Anderson, A.A., Choi, D., Brossard, D., Scheufele, D.A., Liang, X., Ladwig, P.J. and Xenos, M. (2012) 'Coverage of Emerging Technologies: A Comparison between Print and Online Media', *New Media & Society*, 14(6), pp. 1039–1059. doi:10.1177/1461444812439061.

Carter, M. and Egliston, B. (2020) *Ethical Implications of Emerging Mixed Reality Technologies*. Available at: https://ses.library.usyd.edu.au/bitstream/handle/2123/22485/ETHICAL%20IMPLICATIONS.pdf?sequence=2&isAllowed=y (Accessed: 26 October 2020).

Chan, M. (2014) *Virtual Reality: Representations in Contemporary Media*. London: Bloomsbury.

Chuan, C., Tsai, W.S. and Cho, S.Y. (2019) 'Framing Artificial Intelligence in American Newspapers', *Proceedings of the 2019 AAAI/ACM Conference on AI, Ethics, and Society*, Honolulu, USA, 27–28 January, pp. 339–344. doi:10.1145/3306618.3314285.

Chyi, H.I. and Lee, A.M. (2018) 'Commercialization of Technology News', *Journalism Practice*, 12(5), pp. 585–604. doi:10.1080/17512786.2017.1333447.

Cogan, B. (2005) '"Framing Usefulness:" An Examination of Journalistic Coverage of the Personal Computer from 1982–1984', *Southern Journal of Communication*, 70(3), pp. 248–295. doi:10.1080/10417940509373330.

Davies, N. (2009) *Flat Earth News*. London: Vintage.

Davis, F.D. (1989) 'Perceived Usefulness, Perceived Ease of Use, and User Acceptance of Information Technology', *MIS Quarterly*, 13(3), pp. 319–340.

de Vreese, C.H. (2010) 'Framing the Economy: Effects of Journalistic News Frames', in D'Angelo, P. and Kuypers, J.A. (eds.) *Doing News Framing Analysis*. Oxon: Routledge, pp. 187–214.

Dimopoulos, K. and Koulaidis, V. (2002) 'The Socio-Epistemic Constitution of Science and Technology in the Greek Press: An Analysis of its Presentation', *Public Understanding of Science*, 11, pp. 225–241.

Entman, R.M. (1993) 'Framing: Towards Clarification of a Fractured Paradigm', *Journal of Communication*, 43(4), pp. 51–58.

Erjavec, K. (2004) 'Beyond Advertising and Journalism: Hybrid Promotional News Discourse', *Discourse & Society*, 15(5), pp. 553–578.

Evans, L. (2019) *The Re-Emergence of Virtual Reality*. Oxon: Routledge.

Galtung, J. and Ruge, M.H. (1965) 'The Structure of Foreign News: The Presentation of the Congo, Cuba and Cyprus Crises in Four Norwegian Newspapers', *Journal of Peace Research*, 2(1), pp. 64–90. doi:10.1177/002234336500200104.

Gamson, W.A. and Lasch, K.E. (1983) 'The Political Culture of Social Welfare Policy', in Spiro, S.E. and Yutchtman-Yaar, E. (eds.) *Evaluating the Welfare State: Social and Political Perspectives*. San Diego, CA: Academic Press, pp. 397–415.

Gans, H.J. (1980) *Deciding What's News*. London: Constable.

Gitlin, T. (1980) *The Whole World Is Watching*. Berkeley: University of California Press.

Grandinetti, J. and Ecenbarger, C. (2018) 'Imagine Pokémon in the "Real" World: A Deleuzian Approach to PokémonGO and Augmented Reality', *Critical Studies in Media Communication*, 35(5), pp. 440–454. doi:10.1080/15295036.2018.1512751.

Graves, E.K. (2016) *Representations of Virtual Reality in the Media*. Unpublished dissertation. Canterbury Christ Church University.

Graves, E.K. (2017) *Representations of Virtual Reality in UK and US News (2014–2016)*. Masters thesis. Canterbury Christ Church University. Available at: https://repository.canterbury.ac.uk/item/885x2/representations-of-virtual-reality-in-uk-and-us-news-2014-2016 (Accessed: 12 June 2023).

Greengard, S. (2019) *Virtual Reality*. Cambridge, MA: The MIT Press.

Hallahan, K. (1999) 'Seven Models of Framing: Implications for Public Relations', *Journal of Public Relations Research*, 11(3), pp. 205–242. doi:10.1207/s1532754xjprr1103_02.

Hansen, E. (2018) 'The Fourth Estate: The Construction and Place of Silence in the Public Sphere', *Philosophy and Social Criticism*, 44(10), pp. 1071–1089. doi:10.1177/0191453718797991.

Harcup, T. and O'Neill, D. (2001) 'What Is News? Galtung and Ruge Revisited', *Journalism Studies*, 2(2), pp. 261–280. doi:10.1080/14616700118449.

Harcup, T. and O'Neill, D. (2017) 'What Is News? News Values Revisited (Again)', *Journalism Studies*, 18(12), pp. 1470–1488. doi:10.1080/1461670X.2016.1150193.

Harro-Loit, H. and Saks, K. (2006) 'The Diminishing Border between Advertising and Journalism in Estonia', *Journalism Studies*, 7(2), pp. 312–322. doi:10.1080/14616700500533635.

Hetland, P. (2012) 'Internet between Utopia and Dystopia', *Nordicom Review*, 33(2), pp. 3–15.

Kelly, J.P. (2009) 'Not so Revolutionary After All: The Role of Reinforcing Frames in US Magazine Discourse about Microcomputers', *New Media & Society*, 11(1–2), pp. 31–52. doi:10.1177/1461444808100159.

Kim, H., Chan, H.C. and Gupta, S. (2007) 'Value-based Adoption of Mobile Internet: An Empirical Investigation', *Decision Support Systems*, 43(1), pp. 111–126. doi:10.1016/j.dss.2005.05.009.

Kovach, B. and Rosenstiel, T. (2014) *The Elements of Journalism*. 3rd edn. New York: Three Rivers Press.

Lanier, J. (2011). *You Are Not a Gadget*. London: Penguin Books.

Lewis, J., Williams, A. and Franklin, B. (2008) 'A Compromised Fourth Estate?', *Journalism Studies*, 9(1), pp. 1–20. doi:10.1080/14616700701767974.

Li, J., Vinayagamoorthy, V., Schwartz, R., IJsselsteijn, W.A., Shamma, D.A. and Cesar, P. (2020) 'Social VR: A New Medium for Remote Communication and

Collaboration', *Extended Abstracts of the 2020 CHI Conference on Human Factors in Computing Systems*, Honolulu, USA, pp. 1–8. doi:10.1145/3334480.3375160.

Linström, M. and Marais, W. (2012) 'Qualitative News Frame Analysis: A Methodology', *Communitas*, 17, pp. 21–28.

Madary, M. and Metzinger, T.K. (2016) 'Real Virtuality: A Code of Ethical Conduct. Recommendations for Good Scientific Practice and the Consumers of VR-Technology', *Frontiers in Robotics and AI*, 3(3), pp. 1–23. doi:10.3389/frobt.2016.00003.

Maras, S. (2013) *Objectivity in Journalism*. Cambridge: Polity Press.

McKernan, B. (2013) 'The Morality of Play: Video Game Coverage in *The New York Times* from 1980 to 2010', *Games and Culture*, 8(5), pp. 307–329. doi:10.1177/1555412013493133.

McNair, B. (2009) 'Journalism and Democracy', in Wahl-Jorgensen, K. and Hanitzsch, T. (eds.) *The Handbook of Journalism Studies*. Oxon: Routledge, pp. 237–249.

Milgram, P. and Kishino, F. (1994) 'A Taxonomy of Mixed Reality Visual Displays', *IEICE Transactions on Information Systems*, 77(12), pp. 1321–1329.

Moy, P., Tewksbury, D. and Rinke, E.M. (2016) 'Agenda-Setting, Priming, and Framing', in Jensen, K.B., Craig, R.T., Pooley, J.D. and Rothenbuhler, E.W. (eds.) *The International Encyclopedia of Communication Theory and Philosophy*. Chichester: Wiley, pp. 1–13.

Muñoz-Torres, J.R. (2012) 'Truth and Objectivity in Journalism', *Journalism Studies*, 13(4), pp. 566–582. doi:10.1080/1461670X.2012.662401.

National Union of Journalists (2018) *Code of Conduct*. Available at: www.nuj.org.uk/resource/printable-nuj-code-of-conduct.html (Accessed: 8 March 2023).

Pan, Z. and Kosicki, G.M. (1993) 'Framing Analysis: An Approach to News Discourse', *Political Communication*, 10, pp. 55–75.

Pander Maat, H. (2007) 'How Promotional Language in Press Releases Is Dealt with by Journalists', *Journal of Business Communication*, 44(1), pp. 59–95. doi:10.1177/0021943606295780.

Papagiannis, H. (2014) 'Working Towards Defining an Aesthetics of Augmented Reality: A Medium in Transition', *Convergence*, 20(1), pp. 33–40. doi:10.1177/1354856513514333.

Rheingold, H. (1991) *Virtual Reality*. New York: Touchstone.

Richardson, J.E. (2007) *Analysing Newspapers*. Basingstoke: Palgrave.

Rogers, E.M. (2003) *Diffusion of Innovations*. 5th edn. New York: Free Press.

Royal, C. (2006) 'Visualizing Technology: Images in Google and Yahoo News Aggregators', *Proceedings of the 7th International Symposium on Online Journalism*, Austin, Texas, 7–8 April. Available at: http://citeseerx.ist.psu.edu/viewdoc/download?doi=10.1.1.574.4865&rep=rep1&type=pdf (Accessed: 12 February 2018).

Sambrook, R. (2012) *Delivering Trust: Impartiality and Objectivity in the Digital Age*. Available at: https://reutersinstitute.politics.ox.ac.uk/sites/default/files/2017-11/Delivering%20Trust%20Impartiality%20and%20Objectivity%20in%20a%20Digital%20Age.pdf (Accessed: 22 October 2020).

Schäfer, M.S. (2017) 'How Changing Media Structures Are Affecting Science News Coverage', in Jamieson, K.H., Kahan, D. and Scheufele, D. (eds.) *Oxford Handbook on the Science of Science Communication*. New York: Oxford University Press, pp. 51–60.

Scheufele, B.T. and Scheufele, D.A. (2010) 'Of Spreading Activation, Applicability, and Schemas: Conceptual Distinctions and Their Operational Implications for Measuring Frames and Framing Effects', in D'Angelo, P. and Kuypers, J.A. (eds.) *Doing News Framing Analysis*. Oxon: Routledge, pp. 110–134.

Scheufele, D.A. (2013) 'Communicating Science in Social Settings', *PNAS*, 110(3), pp. 14040–14047. doi:10.1073/pnas.1213275110.

Scheufele, D.A. and Lewenstein, B.V. (2005) 'The Public and Nanotechnology: How Citizens Make Sense of Emerging Technologies', *Journal of Nanoparticle Research*, 7, pp. 659–667. doi:10.1007/s11051-005-7526-2.

Shoemaker, P.J. and Reese, S.D. (2014) *Mediating the Message in the 21st Century*. Oxon: Routledge.

Sissons, H. (2012) 'Journalism and Public Relations: A Tale of Two Discourses', *Discourse & Communication*, 6(3), 273–294. doi:10.1177/1750481312452202.

Slater, J.R. (2017) 'Social Constructionism', in Allen, M. (ed.) *The SAGE Encyclopedia of Communication Research Methods*. California: SAGE, pp. 1624–1628.

Sun, S., Zhai, Y., Shen, B. and Chen, Y. (2020) 'Newspaper Coverage of Artificial Intelligence: A Perspective of Emerging Technologies', *Telematics and Infomatics*, 53. doi:10.1016/j.tele.2020.101433.

T3 Online (2015) *Better Than Life: 2015's Hottest VR, Console and PC Gaming Tech*. Available at: www.t3.com/features/the-hottest-gaming-tech-of-2015-so-far (Accessed: 10 March 2015).

Tankard, J.W. (2001) 'The Empirical Approach to the Study of Media Framing', in Reese, S.D., Gandy, O.H. and Grant, A.E. (eds.) *Framing Public Life*. Oxon: Routledge, pp. 95–105.

van Dijk, T.A. (1988) *News as Discourse*. New Jersey: Lawrence Erlbaum Associates.

Whitton, N. and Maclure, M. (2015) 'Video Game Discourses and Implications for Game-based Education', *Discourse: Studies in the Cultural Politics of Education*, 38(4), pp. 1–13. doi:10.1080/01596306.2015.1123222.

Williams, D. (2003) 'The Video Game Lightning Rod', *Information, Communication & Society*, 6(4), pp. 523–550. doi:10.1080/1369118032000163240.

2 Extended Reality Technologies
History and Context

The previous chapter introduced and defined XR technologies and the three types that are of focus in this book (virtual, augmented and mixed reality). This chapter will now provide some more context for these technologies, including their overlapping histories and the current state of the XR market. It then discusses some of the key concepts related to these technologies (namely, immersion and presence), before moving on to explore some of the benefits, concerns and ethical issues surrounding XR. Along those lines, the chapter analyses some of the literature on moral panics, paying particular attention to media panics and technopanics.

History

While virtual and augmented reality have existed for decades, mixed reality is a relatively new concept that developed from AR and VR (Brigham, 2017). Indeed, AR and MR are often grouped together, which will also be the case at certain points within this book. Therefore, this section will provide a brief history of the development of VR and AR specifically, with the understanding that MR's history is intertwined with them both.

Although it is only in recent years that XR for general consumers appears to be gaining traction, VR has appeared in many forms since the 1960s (Steinicke, 2016). Even before this, VR simulators in the military and training situations have been used since the 1930s (Chan, 2014). In 1968, what is believed to be the first VR HMD was created by Ivan Sutherland, named the Ultimate Display but also known as the Sword of Damocles (Rheingold, 1991; Steinicke, 2016). However, such devices were not intended for commercial or mainstream use. In 1986, Jaron Lanier coined the term "virtual reality" (Rheingold, 1991). It was only after this, in the 1990s, when the first attempts at commercial VR were made. This will be referred to in this book as the first wave of XR. Large companies developed and released VR headsets for consumer use, including Nintendo's Virtual Boy and Sony's Glasstron (Ariel, 2017), amongst others. However, these products were not commercially successful (Dixon, 2016), some argue because the technology

DOI: 10.4324/9781003375814-2

was not advanced enough to create a high-quality experience at a reasonable consumer price (Parisi, 2016). Indeed, Rheingold (1991) states that, in 1990, a high-quality VR set-up for one person would cost a minimum of $115,400. Attempting to create these consumer devices at affordable prices meant that the products offered low-quality image resolution, poor ergonomics and physical side effects that could not provide an immersive experience (Ariel, 2017: 36). Therefore, the first wave of XR faded away without success.

As opposed to VR, AR development started slightly later and originated in the workplace environment (Ariel, 2017). In the early 1990s, one of the first AR HMDs was prototyped by Boeing scientists Thomas Caudell and David Mizell to aid the building of aeroplanes (Caudell and Mizell, 1992). Another AR HMD named EyeTap was created in 1999 (Mann, Fung and Moncrieff, 1999), though was never released for consumer use. Also in 1999, HITLab scientists Hirokazu Kato, Mark Billinghurst, Rob Blanding and Richard May developed ARToolKit – an open source software library that enabled the easy development of AR applications (Kato and Billinghurst, 1999). This led to some of the first AR applications being developed for consumer use, including an outdoor mobile game named ARQuake, released in 2000 (Thomas et al., 2000), and MagicBook, which overlaid digital imagery onto books, released in 2001 (Billinghurst, Kato and Poupyrev, 2001). In 2008, the first AR application for a smartphone was released – a travel guide launched with the G1 Android phone (Ariel, 2017). Other applications followed, such as Wikitude and Layar. Therefore, it appears AR has had a more stable presence over the years than VR.

Still, according to Ariel (2017), it was only with the announcement of the AR headset Google Glass in 2012 that the industry started to consider the potential of AR to reach a wide market. Relatedly, it is only recently that the components needed to create a high-quality VR experience have become sufficiently advanced at an affordable price to make consumer VR viable (Parisi, 2016). The announcement and introduction of these products is what this book refers to as the second wave of XR, which is the focus of the study presented here. The Oculus Rift VR headset is considered by many to be the product that spurred this new XR trend (Brigham, 2017; Evans, 2019; Parisi, 2016; Steinicke, 2016). Originally developed by Palmer Luckey of Oculus VR, Oculus Rift gained attention when the company crowd-funded $2.4 million on Kickstarter to create the product, surpassing its initial funding goal of $250,000 in less than 24 hours (Oculus, 2012; Steinicke, 2016). It garnered even more attention in 2014 when the independent company was acquired by Facebook (now Meta) for a value of $2.3 billion (Steinicke, 2016), leading to renewed interest in VR (Brigham, 2017).

However, in the AR realm, Google Glass did not generate any meaningful adoption, leading to it being discontinued in its consumer form in 2015 (Ariel, 2017). Instead, according to Ariel, the smartphone AR game Pokémon Go "marked the beginning of the Augmented Reality Mania"

(2017: 51). The free game was released in July 2016 and, in its first 80 days of release, was installed by over 550 million users and earned $470 million in revenue (Newzoo, 2016). Furthermore, even in 2019, 81 percent of revenue generated by AR mobile games came from Pokémon Go (SuperData, 2020b). Therefore, Pokémon Go has played a significant role in the commercial AR industry so far, at least in the smartphone AR market. In all, it appears that Oculus Rift and Pokémon Go have been major driving forces in the current generation of XR.

State of the Industry

Early estimates predicted that the VR industry would generate approximately $40 billion of revenue worldwide by 2020 (SuperData, 2016). While this did not come to fruition, worldwide XR revenue has increased every year since 2016 and is predicted to continue to do so (Cook et al., 2022; Deloitte, 2018; Markets and Markets, 2023; PwC, 2019; SuperData, 2018, 2020a). According to SuperData (2016), in the year when the first dedicated VR headsets were released to consumers (2016), the industry made approximately $2.8 billion. Though AR/MR revenue was lower than VR in the period this book analyses, AR/MR products have been generating more revenue than VR since 2019 (Deloitte, 2018; P&S Intelligence, 2022; PwC, 2019). This provides useful context because it highlights that VR was more established during the period of this study than AR/MR, while AR and MR have since grown significantly. Moreover, the steady rise in XR revenue, combined with the several large companies involved, suggests that the industry will continue to grow rather than mirroring the commercial failure of the 1990s. Indeed, XR devices are created by companies including Meta/Facebook, Google, Microsoft, HTC, Samsung, Apple, Sony and many more. New devices have continued to be developed and released since this study was carried out, supporting the idea that the technology is now much more established than the first attempt in the 1990s was.

Regarding software, the applications of XR are extremely wide-ranging. Although there are overlaps between VR and AR/MR, the main uses of each vary. During the period this study focuses on (2012–2017), videogames were the main commercial application of VR (Steinicke, 2016). Indeed, according to SuperData (2017), an estimated 65 percent of VR revenue was produced by videogames in 2017. On the other hand, AR/MR did not yet have a main application (Craig, 2013). However, both VR and AR/MR are used in a very wide variety of areas, from entertainment (including videogames and film) to product design and development, training, education, health care, marketing, retail, tourism, defence and more (Ariel, 2017; Blascovich and Bailenson, 2011; Fuchs et al., 2017; Parisi, 2016). Therefore, it is clear that XR is much more than a technology purely for leisure and entertainment.

Immersion and Presence

Immersion and presence are the two key concepts surrounding XR technology, particularly for VR (Brigham, 2017; Evans, 2019). However, these are not new concepts. Broadly, a "stirring narrative in any medium" can create a sense of immersion, defined as the "experience of being transported to an elaborately simulated place" (Murray, 1997: 98). That is to say, individuals could feel immersed in a novel, film or videogame. The concept of presence tends to go hand-in-hand with immersion (Ryan, 2015) and the terms are often used interchangeably (McMahan, 2003). Certainly, Lombard and Ditton's definition of presence has some similarity with Murray's conceptualisation of immersion, with presence described as "the perceptual illusion of nonmediation" (Lombard and Ditton, 1997: n.p.). The illusion of nonmediation "occurs when a person fails to perceive or acknowledge the existence of a medium in his/her communication environment and responds as he/she would if the medium were not there" (Lombard and Ditton, 1997: n.p.). In other words, the user feels as if they are actually present in a simulated environment and instinctively attempts to interact with it as such. It is clear that immersion and presence are strongly linked; based on these definitions, both immersion and presence refer to the illusion of *being* in a simulated world. However, presence expands upon this by referring to the response of an individual. As Lombard and Ditton note above, someone feeling a sense of presence would respond to the simulation as if they were really there. That is to say, with regard to differentiating the two terms, interaction is the key to presence.

Furthermore, immersion and presence have also been appropriated slightly differently in VR technology in comparison to AR/MR technology. In relation to VR, immersion can be understood as "the illusion of being inside a computer-generated scene" (Rheingold, 1991: 112). Put another way, the user feels immersed in a fully digital environment which is different to the physical space they are in. On the other hand, presence in AR (and by extension, MR) "arises from a high level of technologically-facilitated immersion and environmental consistency, and which in turn may give rise to realistic behaviour and response" (Steptoe, Julier and Steed, 2014: 214). With AR and MR, then, immersion is created when the user is convinced that the digital elements are actually present within the physical environment, leading them to interact with the virtual objects as such.

While these concepts have been used in relation to other media before, XR immersion and presence are different in the sense that they are technologically induced rather than being a mental product of the imagination (Ryan, 2015). That is, with the use of a headset or smartphone, this technology provides a sense of immersion and presence by replacing the user's view of the real world with a virtual environment (in VR) or superimposing digital objects onto the real world (in AR/MR). This differs from the sense of immersion that is imagined when reading a novel or watching a film. However, the simple use of an HMD does not guarantee the user will feel immersed or present in the virtual environment. The sense or level of immersion/presence depends upon a number of features, including the quality of the hardware (Steinicke, 2016) and the ability of the user to interact with the virtual environment (Rheingold, 1991). Moreover, Evans argues that immersion is "a tightly crafted emergent property of the visuals, sounds, narratives and haptics (or touch) of the VR experience and the mood or orientation of the user towards the VR experience itself" (2019: 50). Therefore, while immersion and presence are key concepts in this area, it is important to remember that these are not inherent characteristics of any XR experience.

Benefits of XR

There are several benefits of XR, both as a group of technologies and individually. For VR, some argue that immersion can lead to increased empathy towards certain social groups. This idea was introduced in a TED talk by VR filmmaker, Chris Milk, who called VR the "ultimate empathy machine" (Milk, 2015). He argued that immersion can allow users to experience what it feels like to be someone else (for instance, a child refugee), thus leading to increased empathy for such people. In this vein, VR experiences that allow the user to view the world through the perspective of another have been developed. This includes the *New York Times*' 360 degree VR video, *The Displaced* (2017), which shows the story of a child refugee and *Becoming Homeless: A Human Experience* (2017), created by Stanford University's Human Interaction Lab. In addition to content including pre-recorded footage, BeAnother Lab offers a VR experience that allows two users to swap their perspectives to begin to understand what it is like to be in a different body.

However, there is some contention over whether VR can actually make people more empathic. Bollmer posits that "technologies intended to foster empathy merely presume to acknowledge the experience of another, but fail to do so in any meaningful way" (2017: 63). Additionally, Herrera et al.'s (2018) study of VR-induced empathy found that, in the long term (eight weeks after the first stage of the research), the empathy generated by a VR experience simulating homelessness was no greater than the empathy felt by participants who had read a written account of what it was like to be homeless. On the other hand, even after eight weeks, the participants that experienced the VR simulation were more likely than the group that read the written account to have a positive attitude towards the homeless. The VR group was also more likely to take action that could help improve the lives of the homeless, including signing a petition. This indicates that, even if VR may not be the "ultimate empathy machine", as proposed by Milk, it could at least be more effective in bringing about social change than previous methods.

Other benefits surrounding XR are more closely related to the applications of the technology than its immersive capabilities. For instance, XR can be used for pain management (Pourmand et al., 2018) as well as in treating phobias and post-traumatic stress disorder (Greengard, 2019). More specifically, previous studies have found that XR has certain benefits in both education and the industrial sector. For instance, Garzón, Pavón and Baldiris (2019) carried out a meta-analysis of studies examining AR use in education. They found that the technology had several advantages in this area, including the improvement of academic performance, an increase in motivation and an improved understanding of abstract concepts. Similarly, de Souza Cardoso, Mariano and Zorzal (2020) reviewed the literature about AR/MR use in industrial settings (including engineering and manufacturing). Based on these previous studies, they found that some of the main benefits of AR/MR in these areas were improved product quality, reduced workload, improved decision-making and the increased health and safety of workers. Therefore, it appears that AR and MR in particular have notable benefits in these sectors.

Concerns and Ethics

Alternatively, as with most emerging technologies, the introduction of XR brings with it some concerns, both new and old. Currently, most of these concerns have been raised in relation to VR rather than AR or MR due to it offering immersion in a completely virtual world. As Steinicke notes: "The immersive nature of VR raises questions regarding risks and adverse effects that go beyond those aspects in existing media technologies such as smartphones or the Internet" (2016: 145). One of the most salient concerns is VR-induced motion sickness, or cybersickness. Due to the immersive capabilities of VR, "discrepancies between the senses, which provide information about the body's orientation and motion, cause those perceptual conflicts which cannot be naturally handled by the body" (Steinicke, 2016: 47). In other words, immersion causes the user to believe they are moving within a virtual space and, because their actual body is not moving in the same way, this may cause cybersickness. Symptoms can include nausea, headaches, disorientation and vomiting (Evans, 2019; Greengard, 2019; Steinicke, 2016). Much research has been carried out to determine the causes of cybersickness in order to reduce it (for a review see Tian, Lopes and Boulic, 2022). Despite this, cybersickness remains one of the major barriers to the acceptance of consumer VR (Davis, Nesbitt and Nalivaiko, 2015; Evans, 2019; Jeong et al., 2023). Therefore, the attention the news gives to cybersickness could have a significant effect on readers' willingness to adopt XR.

As opposed to cybersickness, some concerns about VR are similar to those associated with videogames. For instance, just as with videogames, there have been concerns over VR users becoming addicted to the virtual experiences the technology provides (Greengard, 2019). Blascovich and Bailenson (2011) even argue that VR could be more addictive than previous media forms due to immersion. A further worry is that this addiction can lead to social isolation or reduce social skills (Greengard, 2019). Also in line with videogame concerns, there are worries surrounding violence in VR experiences. Firstly, being repeatedly exposed to violent scenarios in immersive virtual worlds could lead users to become desensitised to violence (Greengard, 2019). In line with the media violence debate (which will be discussed further below), experiencing this violent content could then encourage users to be violent or aggressive in the real world (Greengard, 2019). Again, the immersive capabilities of VR have made this concern greater with this technology than it has been previously.

Related to violent content, there are additional concerns that some users may experience panic attacks and even strokes or heart attacks when immersed in a disturbing scene (Greengard, 2019). Similarly, distress could be caused to the user if their avatar is assaulted or if their account is hacked (Greengard, 2019; Steinicke, 2016). There are also a range of aliments associated with VR, including eyestrain due to the close proximity of a screen to the face, repetitive strain injury and accidents caused by colliding with real objects while wearing a headset (Greengard, 2019). Other concerns relate to the psychological impact of VR on the user. For instance, some have reported a disillusionment with the real world after experiencing VR (Chan, 2014; Greengard, 2019). Users of VR can also experience depersonalisation-derealisation syndrome, making it difficult to distinguish between the physical and virtual worlds (Steinicke, 2016). Based on these points, it is clear that there are a wide range of concerns when it comes to VR in particular.

For AR and MR, on the other hand, the major concern that is highlighted is privacy (Brigham, 2017; Pase, 2012). This centres around the fact that AR and MR devices can capture or record the physical environment the user is looking at. For instance, when Google Glass was first launched, concerns were raised that confidential information could be recorded without others being aware of it (Brigham, 2017; Greengard, 2019). In addition, this privacy issue links with another concern over surveillance. In order to provide accurate content, AR and MR devices use a mapping technique to monitor where the user is positioned and what they are looking at, leading to fears over the monitoring of the user's location and actions (Carter and Egliston, 2020; Harborth, 2019). Although this discussion of concerns surrounding XR is not exhaustive, these main issues are useful to keep in mind in relation to analysing the media coverage of these technologies.

Moral Panics

Linked to such concerns, the introduction of a new technology is often accompanied by a moral panic (Lim, 2013; Critcher, 2003; Markey and Ferguson, 2017). The concept of the moral panic was first introduced by Stanley Cohen in his PhD thesis which was adapted into his 1972 monograph *Folk Devils and Moral Panics*. This was further developed by Hall et al. (1978) in *Policing the Crisis*. In the third edition of Cohen's book, he describes a period of moral panic as when a "condition, episode, person or group of persons emerges to become defined as a threat to societal values and interests" (2002: 1). Importantly, in the foreword to Critcher's book, *Moral Panics and the Media*, Stuart Allan notes that, despite there being different definitions of a moral panic, one thing they have in common is the idea that "the media play a crucial role in determining the characteristics of a moral panic" (2003: ix). Therefore, analysing news coverage is an important part of examining a moral panic. Hall et al. (1978) highlight four features of a moral panic: (1) concern over a threat is out of proportion; (2) "experts" perceive the threat in similar terms; (3) a threat is shown to have suddenly developed or increased; and (4) the threat is perceived as novel or different from anything before it. Additionally, Goode and Ben-Yehuda (1994) present five characteristics of moral panics, which have some overlap with those set out by Hall et al.: heightened concern; hostility; consensus of threat; disproportionality; and volatility. Thus, if news discourse includes some, or all, of these characteristics, it could be classified as moral panic style coverage.

Moral panics have occurred for a wide variety of topics, from benefit fraud, to migration, to child abuse (Cohen, 2002). However, of particular interest for this book are media panics and technopanics. Drotner (1999) used the term "media panics" to refer to moral panics specifically focusing on forms of media, often new or emerging. Cohen also references media panics, stating that there "is a long history of moral panics about the alleged harmful effects of exposure to popular media and cultural forms" (2002: xix). For instance, Markey and Ferguson (2017: 102) state that the phonograph, radio and television have each been the topic of moral panics across the years. In addition to media panics, there is the more specific "technopanic" introduced by Marwick (2008). As the term suggests, this refers to moral panics focusing on technology. Marwick states that there are three aspects to a technopanic. Firstly, they focus on new media forms, which she classified as "computer-mediated technologies" (2008: n.p.). Secondly, they usually highlight young peoples' use of technology in a negative way. Thirdly, Marwick states: "this cultural anxiety manifests itself in an attempt to modify or regulate young people's behavior, either by controlling young people or the creators or producers of media products" (2008: n.p.). In other words, Marwick argues that technopanics aim to influence the behaviour of young people, perhaps by prompting regulation to be put in place for a new technology.

In her own research, Marwick uncovered that a technopanic was created about the use of online social networking site MySpace. Marwick found that the concerns in this technopanic were mostly about online predators and the privacy of children on MySpace. She concluded that "breathless negative coverage of technology frightens parents, prevents teenagers from learning responsible use, and fuels panics, resulting in misguided or unconstitutional legislation" (2008: n.p.). Indeed, other authors have found that moral, media or technopanics have resulted in regulation being put in place. Dwyer and Stockbridge (1999) uncovered links between moral panics of violent media and several Australian regulatory policies, including ratings systems.

Relatedly, Rogers (2013) argues that the persistent frame of videogame addiction in the news was the reason for it being officially recognised as a condition in the American Psychological Association diagnostic manual in May 2013. This further highlights the importance of examining the news coverage of emerging technologies such as XR to uncover the potential role it could play in prompting regulation.

Regarding other technologies, the mobile phone has also been found to be the topic of a moral panic, or as Goggin termed this – a "mobile panic" (2006: 109). Goggin's 2010 work highlighted the panic over the imaging capabilities of mobile phones linked to sexting. Similarly, in a discourse analysis of Australian newspapers, Jeffery (2018) examined a moral panic about the sexualisation of children that focused on their use of digital technologies, including sexting. Before this, Lemish (2015) discussed the moral panic surrounding teenagers' use of screens. From these examples, the idea that a moral panic often focuses on the impacts on young people seems to hold true.

Moreover, media panics are often associated with the media violence debate, which considers how (or whether) violence in media may induce aggression (Murray, 2013, cited in Piotrowski and Fikkers, 2020). For instance, a moral panic was created surrounding so-called "video nasties" in the 1980s (Petley, 1984). This moral panic focused on "the supposed threat to children posed by their easy access to video cassettes of all kinds" (Petley, 1984: 68). Reports highlighted research by the US National Institute of Mental Health that claimed there was overwhelming evidence linking TV violence to real-life aggression in young people. Petley argues that this was untrue, yet the media presented it "as a given, that there is a direct causal link between violence on screen and violence in real life" (1984: 68). This caused a moral panic to erupt. The news media coverage of video nasties led to films being withdrawn by producers, as well as the creation of new legislation and regulation of these texts.

However, not every new technology will be accompanied by a moral panic. De Keere, Thunnissen and Kuipers (2020) found that this can even be the case if the technology appears to have similarities with others that a moral panic *has* been created about. The authors analysed 681 US news articles about binge-watching (the back-to-back viewing of several episodes of a series using video streaming services such as Netflix). They note that, despite television being the subject of a moral panic and "despite the alarmist label, binge-watching has not sparked a fully-fledged moral panic" (2020: 2). Instead, the news legitimised binge-watching by framing it as manageable and as a high-quality form of entertainment. De Keere, Thunnissen and Kuipers do not suggest any reasons as to why a moral panic has not been created surrounding binge-watching. Nevertheless, the study provides useful insight by showing that moral panics do not always occur when they might be expected to. The next chapter provides further insight into this concept in relation to XR news, alongside an analysis of this news coverage more broadly.

References

Ariel, G. (2017) *Augmenting Alice*. Amsterdam: BIS Publishers.

Billinghurst, M., Kato, H. and Poupyrev, I. (2001) 'The MagicBook: A Transitional AR Interface', *Computers & Graphics*, 25(5), pp. 745–753. doi:10.1016/S0097-8493(01)00117-0.

Blascovich, J. and Bailenson, J. (2011) *Infinite Reality: The Hidden Blueprint of Our Virtual Lives*. New York: Harper Collins.

Bollmer, G. (2017) 'Empathy Machines', *Media International Australia*, 165(1), pp. 63–76. doi:10.1177/1329878X17726794.

Brigham, T.J. (2017) 'Reality Check: Basics of Augmented, Virtual, and Mixed Reality', *Medical Reference Services Quarterly*, 36(2), pp. 171–178. doi:10.1080/02763869.2017.1293987.

Carter, M. and Egliston, B. (2020) *Ethical Implications of Emerging Mixed Reality Technologies*. Available at: https://ses.library.usyd.edu.au/bitstream/handle/2123/22485/ETHICAL%20IMPLICATIONS.pdf?sequence=2&isAllowed=y (Accessed: 26 October 2020).

Caudell, T. and Mizell, D.W. (1992) 'Augmented Reality: An Application of Heads-Up Display Technology to Manual Manufacturing Processes', *Proceedings of the 25th Hawaii International Conference on System Sciences*, Kauai, USA, 7–10 January, pp. 659–669. doi:10.1109/HICSS.1992.183317.

Chan, M. (2014) *Virtual Reality: Representations in Contemporary Media*. London: Bloomsbury.

Cohen, S. (2002) *Folk Devils and Moral Panics*. 3rd edn. Oxon: Routledge.

Cook, A.V., Stanton, B., Arkenberg, C. and Lee, P. (2022) 'Will VR go from niche to mainstream? It all depends on compelling VR content', *Deloitte*, 30 November. Available at: www.deloitte.com/global/en/our-thinking/insights/industry/technology/technology-media-and-telecom-predictions/2023/vr-content-development-lagging-vr-hardware-market.html (Accessed: 3 July 2023).

Craig, A.B. (2013) *Understanding Augmented Reality: Concepts and Applications*. Waltham, MA: Elsevier.

Critcher, C. (2003) *Moral Panics and the Media*. Buckingham: Open University Press.

Davis, S., Nesbitt, K. and Nalivaiko, E. (2015) 'Comparing the Onset of Cybersickness Using the Oculus Rift and Two Virtual Roller Coasters', *Proceedings of the 11th Australasian Conference on Interactive Entertainment (IE 2015)*, Sydney, Australia, 27–30 January, pp. 3–14.

De Keere, K., Thunnissen, E. and Kuipers, G. (2020) 'Defusing Moral Panic: Legitimizing Binge-Watching as Manageable, High-Quality, Middle-Class Hedonism', *Media, Culture & Society*, [Preprint], pp. 1–19. doi:10.1177/0163443720972315.

de Souza Cardoso, L.F., Mariano, F.C.M.Q. and Zorzal, E.R. (2020) 'A Survey of Industrial Augmented Reality', *Computers & Industrial Engineering*, 139. doi:10.1016/j.cie.2019.106159.

Deloitte (2018) *Digital Reality Primer*. Available at: www2.deloitte.com/content/dam/insights/us/articles/4426_Digital-reality-primer/DI_Digital%20Reality_Primer.pdf (Accessed: 3 July 2023).

Dixon, W.W. (2016) 'Slaves of Vision: The Virtual Reality World of Oculus Rift', *Quarterly Review of Film and Video*, 33(6), pp. 501–510. doi:10.1080/10509208.2016.1144018.

Drotner, K. (1999) 'Dangerous Media? Panic Discourses and Dilemmas of Modernity', *Paedagogica Historica*, 35(3), pp. 593–619.

Dwyer, T. and Stockbridge, S. (1999) 'Putting Violence to Work in New Media Policies', *New Media & Society*, 1(2), pp. 227–249.

Evans, L. (2019) *The Re-Emergence of Virtual Reality*. Oxon: Routledge.

Fuchs, P., Guez, J., Hugues, O., Jégo, J., Kemeny, A. and Mestre, D. (2017) *Virtual Reality Headsets — A Theoretical and Pragmatic Approach*. London: CRC Press.

Garzón, J., Pavón, J. and Baldiris, S. (2019) 'Systematic Review and Meta-Analysis of Augmented Reality in Educational Settings', *Virtual Reality*, 23, pp. 447–459. doi:10.1007/s10055-019-00379-9.

Goggin, G. (2006) *Cell Phone Culture: Mobile Technology in Everyday Life*. Oxon: Routledge.

Goode, E. and Ben-Yehuda, N. (1994) 'Moral Panics: Culture, Politics, and Social Construction', *Annual Review of Sociology*, 20, pp. 149–171.

Greengard, S. (2019) *Virtual Reality*. Cambridge, MA: The MIT Press.

Hall, S., Critcher, C., Jefferson, T., Clarke, J. and Roberts, B. (1978) *Policing the Crisis*. London: Macmillan.

Harborth, D. (2019) 'Unfolding Concerns about Augmented Reality Technologies: A Qualitative Analysis of User Perceptions', *Proceedings of the 14th International Conference on Wirtschaftsinformatik*, Siegen, Germany, 24–27 February, pp. 1262–1276.

Herrera, F., Bailenson, J., Weisz, E., Ogle, E. and Zaki, J. (2018) 'Building Long-Term Empathy: A Largescale Comparison of Traditional and Virtual Reality Perspective-Taking', *PLoS ONE*, 13(10). doi:10.1371/journal.pone.0204494.

Jeffery, C.P. (2018) 'Too Sexy Too Soon, or Just Another Moral Panic? Sexualization, Children, and "Technopanics" in the Australian Media 2004–2015', *Feminist Media Studies*, 18(3), pp. 366–380. doi:10.1080/14680777.2017.1367699.

Jeong, D., Paik, S., Noh, Y. and Han, K. (2023) 'MAC: Multimodal, Attention-based Cybersickness Prediction Modeling in Virtual Reality', *Virtual Reality*. doi:10.1007/s10055-023-00804-0.

Kato, H. and Billinghurst, M. (1999) 'Marker Tracking and HMD Calibration for a Video-based Augmented Reality Conferencing System', *Proceedings of the 2nd IEEE and ACM International Workshop on Augmented Reality (IWAR'99)*, San Francisco, CA, USA, 20–21 October, pp. 85–94. doi:10.1109/IWAR.1999.803809.

Lemish, D. (2015) 'Media Moral Panic about Screens and Teens', *Gateway Journalism Review*, 45(339).

Lim, S.S. (2013) 'On Mobile Communication and Youth "Deviance": Beyond Moral, Media and Mobile Panics', *Mobile Media & Communication*, 1(1), pp. 96–101. doi:10.1177/2050157912459503.

Lombard, M. and Ditton, T. (1997) 'At the Heart of It All: The Concept of Presence', *Journal of Computer-Mediated Communication*, 3(2). doi:10.1111/j.1083-6101.1997.tb00072.x.

Mann, S., Fung, J. and Moncrieff, E. (1999) 'EyeTap Technology for Wireless Electronic News Gathering', *Mobile Computing and Communication Review*, 3(4), pp. 19–26.

Markets and Markets (2023) *Extended Reality Industry Worth $111.5 Billion by 2028* [Press Release]. Available at: www.marketsandmarkets.com/PressReleases/extended-reality.asp (Accessed: 3 July 2023).

Markey, P.M. and Ferguson, C.J. (2017) 'Teaching Us to Fear', *American Journal of Play*, 10(1), pp. 99–115.

Marwick, A.E. (2008) 'To Catch a Predator? The MySpace Moral Panic', *First Monday*, 13(6). doi:10.5210/fm.v13i6.2152.

McMahan, A. (2003) 'Immersion, Engagement, and Presence: A Method for Analyzing 3-D Video Games', in Wolf, M.J.P. and Perron, B. (eds.) *The Video Game Theory Reader*. Oxon: Routledge, pp. 67–86.

Milk, C. (2015) *How Virtual Reality Can Create the Ultimate Empathy Machine*. Available at: www.ted.com/talks/chris_milk_how_virtual_reality_can_create_the_u ltimate_empathy_machine (Accessed: 11 March 2020).

Murray, J. (1997) *Hamlet on the Holodeck: The Future of Narrative in Cyberspace*. Cambridge, MA: The MIT Press.

Newzoo (2016) *Analysis of Pokémon GO: A Success Two Decades in the Making*. Available at: https://newzoo.com/insights/articles/analysis-pokemon-go/ (Accessed: 8 March 2020).

Oculus (2012) *Oculus Rift: Step into the Game*. 2 August. Available at: www.kick starter.com/projects/1523379957/oculus-rift-step-into-the-game/posts/279771 (Accessed: 6 March 2020).

P&S Intelligence (2022) *AR and VR Market Report by Type (AR, VR), Offering (Hardware, Software), Device Type (AR Devices, VR Devices), Application (Consumer, Commercial, Enterprise) – Industry Analysis and Growth Forecast to 2030*. Available at: www.psmarketresearch.com/market-analysis/augmented-real ity-and-virtual-reality-market (Accessed: 3 July 2023).

Parisi, T. (2016) *Learning Virtual Reality*. California: O'Reilly.

Pase, S. (2012) 'Ethical Considerations in Augmented Reality Applications', *Proceedings of the International Conference on e-Learning, e-Business, Enterprise Information Systems, and e-Government (EEE)*, Las Vegas, USA, 16–19 July, pp. 38–44.

Petley, J. (1984) 'A Nasty Story', *Screen*, 25(2), pp. 68–75. doi:10.1093/screen/ 25.2.68.

Piotrowski, J.T. and Fikkers, K.M. (2020) 'Media Violence and Aggression', in Oliver, M.B., Raney, A.A. and Bryant, J. (eds.) *Media Effects: Advances in Theory and Research*. 4th edn. Oxon: Routledge, pp. 211–226.

Pourmand, A., Davis, S., Marchak, A., Whiteside, T. and Sikka, N. (2018) 'Virtual Reality as a Clinical Tool for Pain Management', *Current Pain and Headache Reports*, 22(53). doi:10.1007/s11916-018-0708-2.

PwC (2019) *Seeing Is Believing: How Virtual Reality and Augmented Reality Are Transforming Business and the Economy*. Available at: www.pwc.com/gx/en/ technology/publications/assets/how-virtual-reality-and-augmented-reality.pdf (Accessed: 3 July 2023).

Rheingold, H. (1991) *Virtual Reality*. New York: Touchstone.

Rogers, R. (2013) 'Critical Essay — Old Games, Same Concerns', *Technoculture*, 3. Available at: https://tcjournal.org/drupal/vol3/rogers (Accessed: 10 October 2016).

Ryan, M.L. (2015) *Narrative as Virtual Reality 2*. Baltimore: Johns Hopkins University Press.

Steinicke, F. (2016) *Being Really Virtual*. Cham, Switzerland: Springer.

Steptoe, W., Julier, S. and Steed, A. (2014) 'Presence and Discernability in Conventional and Non-Photorealistic Immersive Augmented Reality', *Proceedings of IEEE*

International Symposium on Mixed and Augmented Reality, Munich, Germany, 10–12 September, pp. 213–218.

SuperData (2016) *Virtual Reality to Reach $40B by 2020E*. Available at: www.superd ataresearch.com/blog/virtual-reality-forecast/ (Accessed: 24 March 2016).

SuperData (2017) 'Opportunities in XR: Where Are the Real Opportunities in the Immersive Tech Market?', *VRX 2017*, 7–8 December, San Francisco.

SuperData (2018) *2017 Year in Review: Digital Games and Interactive Media*. Available at: www.superdataresearch.com/market-data/market-brief-year-in-rev iew (Accessed: 30 January 2018).

SuperData (2020a) *SuperData XR Q3 2020 Update*. www.superdataresearch.com/ blog/superdata-xr-update (Accessed: 6 November 2020).

SuperData (2020b) *Year in Review: Digital Games and Interactive Media*. Available at: www.superdataresearch.com/2019-year-in-review (Accessed: 24 February 2020).

Thomas, B., Close, B., Donoghue, J., Squires, J., De Bondi, P., Morris, M. and Piekarski, W. (2000) 'ARQuake: An Outdoor/Indoor Augmented Reality First Person Application', *Proceedings of the 4th International Symposium on Wearable Computers*, Atlanta, USA, 16–17 October, pp. 139–146. doi:10.1109/ ISWC.2000.888480.

Tian, N., Lopes, P. and Boulic, R. (2022) 'A Review of Cybersickness in Head-Mounted Displays: Raising Attention to Individual Susceptibility', *Virtual Reality*, 26, pp. 1409–1441. doi:10.1007/s10055-022-00638-2.

3   An Overview of XR News Coverage

As mentioned in Chapter 1, the study presented in this book utilised both quantitative and qualitative data to examine the media coverage of XR. The quantitative data provides detailed insight into the overall characteristics of the news coverage. It highlights a number of key trends in the discourse, which will be discussed throughout this chapter.

Attention Varies by News Outlet, Year and Technology

Entman (1991) posits that the volume of articles covering a topic indicates the level of importance it is assigned by journalists. With that in mind, measuring the number of articles that were published by the news outlets in this study shows that this level of importance differed by type of XR (virtual, augmented or mixed reality) and news outlet, as well as over time. However, before analysing the data, it is first important to specify how articles were identified from the chosen news outlets (the *Sun*, *Guardian* and *MailOnline*). A search string with two parts was developed. Firstly, the news article had to contain one of three exact terms: "virtual reality", "augmented reality", or "mixed reality". Since these terms are sometimes used to refer to non-XR technology (e.g. the virtual world of a non-XR videogame is sometimes termed "virtual reality"), additional words referring to the headset-based XR this study focuses on were also included in the string. These were as follows: headset(s), helmet(s), goggles, glasses, head-mounted display(s) and hmd(s). Thus, articles were identified that included the term "virtual reality", "augmented reality" or "mixed reality" as well as any one of the terms referring to headsets above. As mentioned previously, this resulted in a final sample of 977 news articles.

By far, the *MailOnline* published the most news articles about XR, with 668 articles making up 68 percent of the total sample. The *Guardian*, though much less than the *MailOnline*, still published a substantial number of articles that met the sample criteria (248). Alternatively, the *Sun* paid very little attention to XR at all, with just 61 articles appearing on their website that met the above criteria. Moreover, the *Sun* did not publish any articles about

DOI: 10.4324/9781003375814-3

XR that met the criteria until 2015, when just one article was found. It was only in 2016 that this news outlet started paying more attention to XR, though even this figure was low compared to the other news outlets (24 articles). On the other hand, the *Sun* was the only news outlet to publish more articles about XR in 2017 than 2016, which suggests this publication judged the newsworthiness of XR differently to the *MailOnline* and the *Guardian*.

There could be a number of reasons for this result. Firstly, the owners of news outlets can affect how their outlets present topics in news articles (Witschge, Fenton and Freedman, 2010). Through interviews with journalists, Kovach and Rosenstiel found that the content of a news outlet was "heavily influenced by the values of the ownership" (2014: 285). In 2016, the Australian arm of the *Sun*'s owner, News Corp, was the lead investor in the first funding round of AR company Plattar (Bennett, 2016). Additionally, as of 2017, News Corp was the largest shareholder of VR company PropTiger (News Corp, 2017). This could have prompted the *Sun* to start reporting on XR in 2016. However, the *Sun* still did not publish a high volume of articles on its website about XR in any year studied, meaning this is likely not the cause. Alternatively, it may be that XR was simply not high on the *Sun*'s agenda due to the typical focus of tabloid news outlets on sensationalist and celebrity news (Zelizer and Allan, 2010) rather than subjects such as technology and science. Nevertheless, it appears that the *Guardian* and the *MailOnline* attributed much more importance to XR during its inception than the *Sun* did. This indicates that, in relation to the hierarchy of influences model (Shoemaker and Reese, 2014), the media organisation factor has had an impact on the frame-building process of XR news as the news outlets paid varying attention to the technology.

This attention also varied over time. Chapter 1 highlighted that 2012 was chosen as the first year of the sample period because this was the year the first products of the recent XR trend were publicly announced (such as Google Glass and Oculus Rift). Despite this, only 24 news articles were published about XR in 2012, increasing very slightly to 33 in 2013 (see Figure 3.1). Therefore, the initial announcements of Google Glass and Oculus Rift do not appear to have garnered much attention from the news outlets. On the other hand, this figure increased substantially in 2014 when 122 articles were found using the search criteria. In 2014, a major event for the XR industry was Facebook (now Meta) owner Mark Zuckerberg acquiring the independent VR company Oculus for $2.3 billion (Steinicke, 2016). The increase of news articles in this year suggests that Facebook's involvement in the XR industry made the technology appear more newsworthy to journalists, thus resulting in increased coverage.

This relates to the concept of news values. News values are factors that are used by journalists to determine whether or not something is worthy of becoming a news item (Bednarek and Caple, 2012). Starting with Galtung and Ruge (1965), many other authors have adapted, developed and added to these news values (see, for example, Bell, 1991; Caple and Bednarek, 2013;

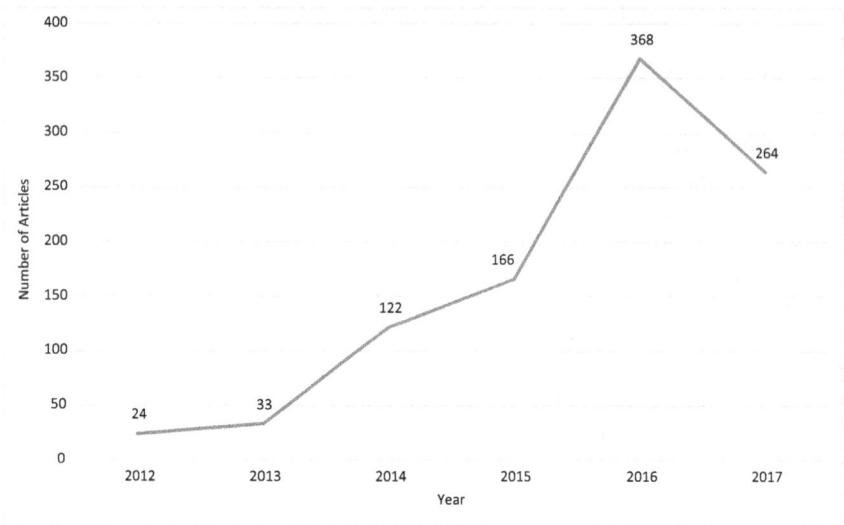

Figure 3.1 Number of XR news articles published per year across the *Sun*, *Guardian* and *MailOnline*.

van Dijk, 1988; Gans, 1980; Harcup and O'Neill, 2001, 2017). The growth in XR news coverage when Zuckerberg became involved indicates that the power elite news value (Harcup and O'Neill, 2017) has impacted the amount of attention the news afforded XR. This news value suggests that journalists are more likely to report on a news item if it is related to people with considerable power or popularity within society. Within the frame-building process, the hierarchy of influences model considers news values as routine practices (Shoemaker and Reese, 2014), indicating that this factor has affected the framing of XR, at least at the basic level of attention given to the topic.

While 2014 marked the first year XR was given substantial attention by the news, the number of articles written about the technology peaked in 2016. In 2016, 368 articles were published across the news outlets: over twice as many as the previous year. Often dubbed as "the year of virtual reality" (Fuchs et al., 2017; Steinicke, 2016), 2016 saw the release of Oculus Rift as well as other dedicated VR products, including HTC Vive and PlayStation VR. This means that 2016 would have been the year that early adopters made their decision of whether to purchase a VR product. Entman argues that when many reports are published about a topic, they "may penetrate the consciousness of a mass public" (Entman, 1991: 9). Therefore, the fact that a large number of news articles were published about XR in 2016 could have increased the awareness of XR, thus supporting the first stage of the innovation-decision process (knowledge-building (Rogers, 2003)). In other words, through paying particular attention to XR news in this year, these news outlets have supported its adoption by increasing the potential consumer base for the technology.

Lastly, there were clear differences between XR type as well. Overall, 61 different XR headsets were mentioned, with 41 of these being VR devices, 17 AR and three MR. Additionally, 63 percent of articles cited VR products, whereas 14 percent mentioned AR devices and 11 percent named MR products. These initial figures show that the news was much more likely to report on VR than AR or MR. Moreover, although 61 different devices were mentioned at least once, only nine of these appeared in ten or more articles. Out of these nine, six were VR headsets, one was an AR device and two were MR headsets, further emphasising the focus on VR. This was fairly consistent per news outlet, showing that within the hierarchy of influences model (Shoemaker and Reese, 2014), the media organisation reporting on XR has not had a significant impact on the type of XR given the most attention.

Several VR products were released for general consumer use during the sample period of this study, whereas AR and MR products were mostly targeted towards developers of these platforms. Since the news outlets that were examined are targeted towards the general public, the choice to focus on VR displays the relevance news value which suggests that "information is preferred about events or actions that are relevant for the reader" (van Dijk, 1988: 122; see also Harcup and O'Neill, 2017). Products that are targeted towards the same audience as the news outlet are arguably of higher relevance to their readers than products that are targeted towards a more specialised audience. This could explain the focus on VR as opposed to AR or MR. As explained above, news values relate to the routine practices factor of the frame-building process (Shoemaker and Reese, 2014), thus highlighting another instance of this factor impacting the framing of XR.

Looking at specific products, Oculus Rift was cited in the largest portion of articles overall (nearly 50 percent) and significantly more than others (see Table 3.1). Every news outlet mentioned this product the most, demonstrating that the headset that is thought of as being the one to start the new XR trend (Steinicke, 2016) was a major focus for all publications. Therefore, it appears that the news outlets have attributed high importance to Oculus Rift in particular (and more so) than any other XR product. Furthermore, every device that was mentioned in ten or more articles was created by, or had connections with, a large technology company (Facebook, Samsung, HTC, Sony, Google and Microsoft). Thus, as well as the relevance news value, the power elite news value (Harcup and O'Neill, 2017) also appears to have played a role in which devices were framed as important by the news outlets.

However, it is worth noting that not every year focused on VR. In 2012, AR/MR products (grouped together due to their similarities and because so few MR devices were mentioned) appeared in 50 percent of articles, while VR devices only appeared in 4 percent of articles. Additionally, in 2013, 70 percent of articles mentioned AR/MR devices, in comparison with the

Table 3.1 Number of articles mentioning top devices per news outlet

Device (type)	Sun		Guardian		MailOnline		Overall	
	No.	Percent	No.	Percent	No.	Percent	No.	Percent
Oculus Rift (VR)	17	42.50	147	75.77	321	64.46	485	49.64
Samsung Gear VR (VR)	14	35.00	55	28.35	143	28.71	212	21.70
HTC Vive (VR)	9	22.50	57	29.38	133	26.71	199	20.37
PlayStation VR (VR)	9	22.50	62	31.96	123	24.70	194	19.86
Google Cardboard (VR)	5	12.50	50	25.77	85	17.07	140	14.33
Google Glass (AR)	3	7.50	36	18.56	88	17.67	127	13.00
Microsoft HoloLens (MR)	3	7.50	20	10.31	70	14.06	93	9.52
Magic Leap (MR)	0	0.00	12	6.19	28	5.62	40	4.09
Google Daydream View (VR)	2	5.00	8	4.12	28	5.62	38	3.89

24 percent that included VR products. Still, it should be remembered that the number of news articles published in these years was relatively low, meaning that their importance was not highlighted strongly. In 2014, when Facebook acquired Oculus and the news outlets began publishing more articles on XR in general, VR products became the focus of the coverage. From then onward, VR products were mentioned significantly more than AR/MR devices every year. This is further evidence to suggest that Facebook's involvement with XR had a substantial impact on the focus of the coverage, as was indicated by the sharp increase in number of articles published in 2014.

Products and Applications as Key Topics

By recording the main topic of each article, the content analysis revealed that the most common topic overall was "application(s)", which was the main focus in 49 percent of articles (see Appendix 1). In other words, the news outlets paid the most attention to XR software. Second to this, "product(s)" was the main topic of 25 percent of articles, showing that XR hardware was also the focus of a significant number of reports. Aside from applications and products, the third most common topic was "demo", though this was only the main focus of 6 percent of articles. This makes clear just how much the "application(s)" and "product(s)" topics dominated the coverage. Indeed, these two topics were the most common every year in each news outlet, although there was some fluctuation in which was the most used out of the two. Additionally, there were no substantial differences between the topics of VR articles in comparison to AR/MR. Therefore, the vast majority of articles focused on describing the features and uses of XR regardless of news outlet, year or XR type, with a particular emphasis on software. Furthermore, as well as applications being the focus of approximately half of the articles, almost all articles (94 percent) mentioned at least one use of XR. This shows that it was extremely common for news articles to note the uses of XR, whether it was the main focus or not.

Rogers argues that, during the first stage of the innovation-decision process (knowledge-building), "an individual mainly seeks software information" about an innovation (2003: 21). Therefore, this focus on the applications (i.e. software) of XR could support the diffusion of the technology by providing potential adopters with the information most relevant to them in these early stages. Additionally, focusing on products and applications increases the observability of XR as an innovation. According to Rogers, the observability attribute "is the degree to which the results of an innovation are visible to others" (2003: 16). Publishing news articles specifically focusing on these products and applications does just that, suggesting that this could support the diffusion of XR.

This is very different from moral panic style coverage, which is typically characterised by exaggerated concerns or fears as well as calls for regulation (Cohen, 2002; Marwick, 2008; see Chapter 2). Indeed, it is significant that only 29 articles (3 percent) focused on concerns surrounding XR and "regulation" was the least common topic, being the focus of just two articles. These results show that news representations of XR may differ from those of other technologies that have been found to be the subject of a moral panic, such as radio, TV (Markey and Ferguson, 2017), mobile phones (Goggin, 2006) and videogames (Rogers, 2013). To the other extreme, the lack of articles about concerns and regulation suggests that the news may not be paying *enough* attention to the potential risks and negative implications of this technology that the public should be aware of when deciding whether to adopt XR.

Instead, the focus on applications coincides with the findings presented by Dimopoulos and Koulaidis (2002) about science and technology in the Greek press, Cogan's (2005) study of the PC in the US and Kelly's (2009) analysis of microcomputers in US magazines, which all uncovered a focus on the uses of these technologies. Moreover, the fact that the vast majority of articles pay the most attention to XR applications and products suggests that the content of these articles may have more in common with lifestyle journalism than traditional news. To expand, Hanusch defines lifestyle journalism as providing audiences with information "about goods and services they can use in their daily lives" (2012: 2). This is what articles about products and applications do, showing clear similarities with this genre of journalism.

Despite this, XR news is still generally presented to consumers as "news". Indeed, part of framing involves not only what is said about a topic but also how and where the article appears, which can set the tone for what the audience is reading (Tankard, 2001). Thus, the way articles are labelled by news outlets can impact the way they are perceived by audiences. In this sample, the majority of articles were labelled as "news" on the news sites (86 percent), while the remaining were labelled with other variables such as feature and opinion. The journalistic norm of labelling articles that are not purely news "reinforce[s] the legitimacy and authority of the other news stories as being factual" (Pan and Kosicki, 1993: 62). Thus, for the majority of articles in this sample to be presented as news means they will appear to be based on factual information rather than opinions. Certainly, Pan and Kosicki expand that this increases the truthful value of the news, as well as the likelihood that the audience will accept these frames (1993: 62). Therefore, this type of labelling could make the frames applied to XR in these news articles more persuasive, which is of concern if the discourse has more in common with lifestyle journalism that is known for treating audiences as consumers.

An Entertainment Technology

Examining which application types were mentioned provides more insight regarding this consumer focus. Although XR has a wide variety of applications, the news outlets in this study primarily focused on entertainment and leisure uses. Out of 41 application types, videogames were mentioned in the most articles (47 percent) and much more than any other application type (see Appendix 2). Second to this, film/TV/video applications were mentioned in 19 percent of articles overall. Both of these application types involve using XR for entertainment or leisure, which could frame it as a technology to be used for fun, rather than a technology with the potential to have a significant impact on society. How a technology can or should be used is defined in its emergence by discursive outlets (McKernan, 2013), such as the news. Therefore, focusing on leisure applications could encourage readers to perceive XR as an entertainment medium.

While all news outlets mentioned videogames the most, there was some variation between them regarding the other applications that were mentioned. The *Guardian* and *MailOnline* were fairly similar in which applications they focused on, with each citing videogames, film/TV/video and social media/communication uses the most. On the other hand, the *Sun* differed quite drastically. The second most mentioned application type in the *Sun* was pornography, teledildonics and sex (23 percent), followed by theme park and rides (13 percent). These application types rarely appeared in the other outlets. Traditionally, tabloid news outlets such as the *Sun* are expected to put more emphasis on sensationalist and entertainment news styles than quality news outlets (Zelizer and Allan, 2010), which could be the reason for this difference. Indeed, tabloids are also known for their focus on sex (Carvalho and

Burgess, 2005), which explains the extra attention paid to pornography, teledildonics and sex uses in the *Sun*. Therefore, it appears that, in terms of applications, the *Sun* has framed XR slightly differently than the *Guardian* and *MailOnline*, albeit still focusing on entertainment or leisure uses.

More serious applications were much less common. Overall, the serious application types mentioned the most were health (appearing in 12 percent of articles) and education (appearing in 11 percent of articles). However, the *Sun* rarely mentioned health or education applications. The *MailOnline* mentioned health uses in 11 percent of articles and education in 9 percent, while the *Guardian* was most likely to include references to health applications (18 percent of articles) and education uses (17 percent of articles). Furthermore, the *Guardian* also noted social change and awareness applications in 10 percent of articles – substantially more than the *MailOnline* and the *Sun*. Therefore, the *Guardian* appears to have made more attempts at framing XR as a serious technology compared to the other news outlets. Since quality news outlets are typically expected to offer more sober reporting than middle-market or tabloid outlets (Bastos, 2019), these results offer some support for this claim in relation to XR news. Overall, it appears that the media organisation factor (Shoemaker and Reese, 2014) has impacted the frame-building process of XR news in terms of representing how the technology can be used.

In a different way, it is also significant that health and education uses were the most commonly mentioned serious XR applications. Other serious uses were rarely referenced by the news articles, such as training, military and defence and architecture/planning. This shows that, when the news coverage *has* mentioned serious applications, it is the kind that impacts the majority of the population (education and health) rather than more niche areas. Thus, it appears that the relevance news value (van Dijk, 1988; Harcup and O'Neill, 2017) has again played a role in the frame-building process in XR articles, showing another way in which routine practices factor (Shoemaker and Reese, 2014) has impacted the framing of XR.

As well as variations between news outlets, there were some notable differences between the applications referred to in the different years of the sample. In 2012, social media and communication uses were mentioned the most (25 percent of articles). Additionally, in 2013, the most cited application type was photography/video recording (39 percent). Other uses that appeared in a large portion of articles in 2012 and 2013 were tourism/travel (which in these years referred to map navigation) and web browsing. On the other hand, since 2014, videogame applications were mentioned the most every year. Considering this data alongside a comparison between VR articles and AR/MR articles suggests that this change is due to a shift in focus on XR type. To expand, although videogame applications appeared the most in VR articles (49 percent), articles about AR/MR mentioned social media and communication uses as often as they did videogames (both appearing in 28 percent of articles). Furthermore, although film/TV/video uses were the second most common in VR articles, this application type only appeared in 3 percent of AR/MR articles. Aside from videogames, AR/MR articles were more likely to focus on uses that are typically associated with smartphones, such as photography/video recording, web browsing and tourism/travel. Therefore, VR seems to have been framed slightly differently to AR/MR in terms of its uses.

The focus on smartphone-related uses for AR/MR products means these articles highlight the compatibility attribute of the innovation. According to Rogers, compatibility "is the degree to which an innovation is perceived as being consistent with the existing values, past experiences, and needs of potential adopters" (2003: 15). As smartphones are familiar to a wide portion of the population (particularly those accessing online news), mentioning applications that are related to smartphones presents AR/MR as highly compatible with the past experiences and needs of potential adopters. The same can be said for the focus on videogames and film/TV/video applications in VR coverage, since these are leisure activities a large portion of the general public are familiar with. While the focus on XR as a technology for entertainment or leisure does not emphasise its importance as much as if serious uses were referenced the most, it does mean that the product may appeal to a wider audience. Furthermore, the entertainment focus links to Kim, Chan and Gupta's value-based adoption model which suggests that experiencing "immediate pleasure or joy from using a technology" makes it more likely to be adopted (2007: 116). Therefore, framing the uses of XR in this way could potentially support the diffusion of the technology.

The Influence of XR Creators

The volume of articles published about XR in different years is already some indication that owners of these products have impacted the news coverage. Supporting this further, an analysis of news sources shows that such groups played a key role in shaping the discourse. Application creators and device creators were the most used sources in XR news in terms of quotations and paraphrased statements as well as multimedia content. Overall, sources falling into the category of application creators (referring to individuals or organisations creating XR software) appeared in 40 percent of articles. Similarly, quotes or paraphrased statements from device creators (referring to individuals or organisations producing XR hardware, such as headsets or peripherals) appeared in 39 percent of articles. These results show that, not only have applications and products been the main focus of articles, but the companies and individuals creating this software and hardware have been the groups most able to shape the discourse.

Who is allowed to speak within a news article can determine which individuals or groups become the primary definers of a topic (Hall et al., 1978). For instance, Critcher states that the "media act as secondary definers whose function is to reproduce the definitions of primary definers and, in the

popular press especially, to 'translate' official statements into everyday language" (2003: 134). Here, XR creators have become the primary definers of
XR, meaning their voices have been particularly prominent in the framing
of the technology. Since such sources are invested in the success of XR, this
means that the publications have afforded substantial power to voices that
are unlikely to be negative or critical about XR. If journalists use these types
of sources the most when producing XR news, it is unsurprising that concerns
and regulation were rarely the main focus of articles. Certainly, moral panic
style coverage would directly conflict with the interests of these sources that
are aiming to sell XR hardware and software. Source choice comes under
the social institutions factor of the hierarchy of influences model (Shoemaker
and Reese, 2014), showing how this factor has impacted the frame-building
process in XR news.

Still, there were some slight differences between the news outlets regarding
which sources were used the most (see Appendix 3). Whereas application
and device creators were most often used as sources in the *Guardian* and
MailOnline, the two most used sources in the *Sun* were application creators
(28 percent) and general users (as opposed to professional users), which were
quoted or paraphrased in 18 percent of articles. Device creators were still
the third most used source in the *Sun* (16 percent), though they appeared
significantly less than in the *Guardian* and *MailOnline*. The reason for this
variation could be that the *Sun* only started substantially reporting on XR
in 2016 when more XR products had been released to consumers and, thus,
more people were able to use them and share their experiences. Additionally,
tabloid news outlets such as the *Sun* are known to focus more on human
interest stories than quality news outlets (Bird, 2009), perhaps explaining
this publication's preference to use the general public as sources. Regardless
of the reasoning, this shows that the *Sun* has been less reliant on elite
sources (such as application and device creators) than the *Guardian* and the
MailOnline in the process of framing XR. The use of elite sources can make
discourse more persuasive (van Dijk, 1988: 87), meaning that the framing of
XR in the *Guardian* and *MailOnline* could have a stronger impact than the
frames that appear in the *Sun*. Furthermore, this shows that the media organisation factor of the frame-building process (Shoemaker and Reese, 2014)
has affected whose voices are heard to some degree, although the differences
were not very stark.

A similar result can be identified when considering multimedia attribution.
Out of the 4,642 multimedia items used in the news articles (e.g. images,
videos, GIFs), the largest portion originated from device creators (17 percent exactly; see Appendix 4). The second most common attribution type
was news agencies (almost 17 percent) and application creators ranked third
(14 percent). There were no notable differences between XR type or news
outlet regarding the sourcing of multimedia. Further to this, device creators
were the most used source for multimedia in every year of the sample, until
2017 when application creators became the most used. It is clear that these

sources were consistently used across the period studied. This shows that creators of XR hardware and software have not only been able to define XR in their own words through quotations, but have also been able to define XR visually through the inclusion of their own imagery and videos. As not all sources will frame a topic in the same way, framing contests may occur (Hallahan, 1999). Gamson states that framing contests "highlight the central importance of the relationship between journalists and sources and the process of selecting sources to quote" (2001: ix). Certainly, the selection of sources in XR news has prioritised the groups that are invested in the success of XR, and, thus, are advocates of positive frames. By prioritising these voices, the news media avoids critical comments about XR (and, indeed, moral panic style coverage), instead focusing on those sources that would frame XR in a positive light.

As will be explored further below, the commercial pressures on newsrooms mean that journalists are increasingly expected to produce large volumes of content as quickly as possible, which can lead to a reliance on easily accessible sources (Lewis, Williams and Franklin, 2008). This is amplified in online news, where the time between gathering information and publishing an article can be "a matter of minutes" (Forde and Johnston, 2013: 115). In interviews with UK technology journalists, Brennen, Howard and Nielsen (2020) found that these time shortages were problematic because journalists had a lack of resources to both understand and translate complex technological issues for the audience. This led to relying on industry insiders (such as company announcements) and other media outlets as sources. Such practices could explain why these sources dominate XR news coverage, thereby providing XR companies with the power to define the technology to the general public in a way that benefits their commercial interests. Within the hierarchy of influences model (Shoemaker and Reese, 2014), the social system of capitalism that these news outlets operate in appears to have impacted the frame-building process by affecting which sources are included in the news content.

Native Advertising

Related to the impact of the capitalist social system is the practice of native advertising. Native advertising can be understood as "a form of paid content marketing, where the commercial content is delivered adopting the form and function of editorial content" (Conill, 2016: 905). Native advertising is strongly linked to Fairclough's concept of marketisation which suggests that "advertising and promotional discourse have colonized many new domains of life in contemporary societies" (1993: 139). This can lead to the discourse of consumerism, typically found in advertising, "colonizing" other areas (Fairclough, 1989: 209). While Fairclough originally used this concept to refer to texts such as university prospectuses, the practice of native advertising certainly suggests this is relevant for news content as well. Indeed, several studies have analysed the supposed blurring of news and promotional content by examining native advertising. For instance, in the context of Slovenia, Erjavec (2004) found that promotional news reports (i.e. texts that have been paid for but are presented as news) "were in no way separated from the editorial content by layout, position or labelling" (2004: 562). Similarly, the structure of the promotional news reports was no different from others that had not been paid for. This means that readers would not be able to determine that the article had been paid for by appearance alone, thus being an example of native advertising. In line with Erjavec, Harro-Loit and Saks' (2006) study of Estonian media found that these hidden advertisements were made to look and read like a standard editorial news report, "thereby making it difficult for the reader to recognise it as promotional material" (2006: 317). This lack of transparency certainly does not have readers' best interests in mind.

Without researching journalists at work, another way in which native advertising can be identified is through the use of hyperlinks to retailers (Wojdynski, 2016). In relation to XR, 8 percent of links in articles directed the reader to a retail site where they could purchase XR hardware or software. In addition, the use of links to retailers peaked in 2016 at 11 percent. Since 2016 was the year several VR products were commercially released, this suggests that the news articles have encouraged the adoption of XR during this crucial year by directing traffic towards sites where readers could buy those products. This is supported by the fact that links to retailers were more common in VR articles than they were in AR/MR articles, since more VR products were commercially available at the time.

Moreover, while these are not high figures, the use of any links to retailers at all hints that these news outlets may have some financial incentive for framing XR positively. Indeed, 11 articles in the *Guardian* included the following statement:

This article contains affiliate links, which means we may earn a small commission if a reader clicks through and makes a purchase. All our journalism is independent and is in no way influenced by any advertiser or commercial initiative.

Although the *Guardian* stresses that they are not influenced by commercial forces, the appearance of links to retailers raises doubts about this claim. Certainly, out of all news outlets, the *Guardian* was most likely to include links to retailers (identified in 10 percent of articles). As the quality news outlet in this sample, the fact that the *Guardian* did this most often is surprising, since it would be expected for them to follow journalistic standards (such as the independence they cite) more strictly (Bastos, 2019). Still, the *Sun* and *MailOnline* both included some links to retailers as well (in 5 and 4 percent of articles respectively), showing that this practice was not exclusive to the *Guardian*. Each news outlet has included at least some links to retailers, suggesting native advertising may be present within these articles.

Embedding links was not the only way the news articles directed readers towards locations they could purchase XR products. Overall, 10 percent of articles included details about how or where XR products could be purchased. Both the *Guardian* and the *Sun* included such information in similar portions of their articles (approximately 15 percent each), whereas 8 percent of *MailOnline* articles did this. Related to lifestyle journalism, Arik and Çağlar (2005) also identified articles mentioning where a product could be purchased. According to the authors, such coverage supports consumption culture. Indeed, this is one way that the news outlets have *directly* supported XR diffusion. This compromises the independence of the press (Lewis, Williams and Franklin, 2008) and benefits the XR companies aiming to sell these devices and applications. Moreover, while these figures are fairly small, the fact that any articles included this information is further indication that native advertising is present.

There is a direct financial relationship between media outlets and their advertisers that can impact the frame-building process. News outlets rely on advertising as their main form of income (Bednarek and Caple, 2012; Bettig and Hall, 2012), meaning that advertisers can influence journalists to run/ not run a story or to write about a topic in a certain way (Bettig and Hall, 2012; McManus, 1995). Advertisers "have considerable power to dictate favorable public messages" (Shoemaker and Reese, 2014: 116), whether directly or indirectly. Regarding direct influence, advertisers have been known to retract support if their products or services have been written about critically by an outlet (Bettig and Hall, 2012). This links to the indirect influence that can occur as a result of journalists attempting to prevent such issues. That is to say, news outlets may be less likely to include critical coverage of their advertisers' products and services in order to avoid backlash from those advertisers. The presence of native advertising in XR news highlights a financial relationship between the outlets and XR companies. If these publications gain some financial reward for getting readers to purchase an XR product, this could impact the frame-building process by encouraging positive framing of XR.

Furthermore, aside from the use of hyperlinks to retailers, it is clear that these news outlets have relationships with XR practitioners since some news articles were even written by creators of XR applications. Overall, six such articles appeared in the news outlets. The majority of these (five) appeared in the *Guardian*, while the *MailOnline* published one such article. On the topic of health reporting, Lipworth et al. note that "relationships between companies and journalists may impact negatively upon journalistic principles such as integrity and fairness" (2015: 252). These effects could extend to other areas of journalism, such as technology news. To maintain their connections, journalists writing for these publications may also avoid writing critically about XR. This provides insight into how the social institutions factor of the hierarchy of influences (Shoemaker and Reese, 2014) could impact the frame-building of XR regarding the relationships between news organisations and other stakeholders.

Overall, it appears that XR news has been commercialised. According to McManus, news commercialisation can be defined as *"any action intended to boost profit that interferes with a journalist's or news organization's best effort to maximise public understanding of those issues and events that shape the community they claim to serve"* (2009: 219, original emphasis). As McManus highlights: while the norm of journalism "is to inform the public", the *"norm of business* is to maximise profits over an indefinite period" (McManus, 1995: 308, original emphasis). McManus argues that commercialisation has led journalism to prioritise the norms of business. Such news is problematic because it "leaves little room for ethics, professionalism, objectivity and the things that constitute *journalism"* (Richardson, 2007: 79, original emphasis). In the case of XR news, the capitalist social system that news organisations operate within appears to have had an impact on the way that XR has been framed in the news.

Commercial Pressures and Journalistic Integrity

Further evidence of this impact can be seen elsewhere. To finish off this chapter, it is worth mentioning some results that say less about how XR has been framed but provide some indication that journalists may be under pressure when creating this news content. Since news readership shifted online and away from print, news outlets have faced a reduction in revenue. Advertising, news organisations' largest income-generator (Bettig and Hall, 2012), is much less profitable online than it is in print (Rosenkranz, 2016; Williams and Clifford, 2009). This loss of revenue has resulted in staff cuts alongside higher demand for news content which means that journalists have much heavier workloads but no additional time (Williams and Clifford, 2009: 18). Within XR news, there are multiple indications that these time constraints have impacted the quality of the coverage.

The first is the use of news wire or agency copy. Overall, according to article bylines, 15 percent of articles were written by news agencies, with the majority of these appearing in the *MailOnline*. Only one article in the *Sun* and two in the *Guardian* had agency bylines, whereas 22 percent of *MailOnline* articles were written solely by news agencies (i.e. with no changes made before publishing). While it is possible that the *Sun* and the *Guardian* journalists integrated agency copy into their reports without specifying this (as was found in Lewis, Williams and Franklin's (2008) study), what is known for certain is that the *MailOnline* has been particularly reliant on agency material when reporting on XR. Moreover, 17 percent of multimedia content was attributed to news agencies. This ranged from 14 percent of multimedia in the *Sun* to 19 percent in the *Guardian*, with the *MailOnline* in the middle at 17 percent. These results suggest that the *Sun* and the *Guardian* have had more input from news agencies than was initially implied by the article bylines, since multimedia are usually accompanied by a press release.

In terms of the hierarchy of influences model (Shoemaker and Reese, 2014), this shows that the routine practice of the *MailOnline* to publish unedited agency copy has played a role in the frame-building process, allowing these agencies to have substantial opportunities to frame XR through this outlet. It also indicates the practice of "churnalism", which involves journalists "recycling second-hand wire copy and PR material without performing the 'everyday practices' of their trade" (Davies, 2009: 59). This could be a result of the increasing commercial pressures on journalists. Thus, the capitalist social system that these news organisations operate within has impacted the frame-building process, resulting in power being afforded to news agencies to define XR. Since this agency material is sent to many other news outlets that may either adapt or publish the copy verbatim (Lewis et al., 2008), this could result in a lack of diversity in viewpoints and topics in XR news (which will be discussed in later chapters).

Moreover, it is best practice in journalism to identify sources (whether of quotes or multimedia content) to maintain transparency and accuracy (Bull, 2010). However, a substantial portion of multimedia did not have an attribution at all (21 percent). In the *Sun*, 30 percent of multimedia were missing an attribution, 23 percent were unattributed in the *MailOnline* and, even in the *Guardian*, 8 percent had no attribution. Thus, while it was more common for multimedia to be unattributed in the tabloid and mid-market outlets, the quality publication also had this flaw. A similar practice can be observed regarding the use of sources for quotes and paraphrased statements. Overall, 12 percent of articles included statements without listing their source and every news outlet had this flaw to some extent, indicating a lack of journalistic integrity regarding source attribution. This is further evidence to suggest that the commercial pressures of the newsroom have caused a reduction in news quality, as was also indicated by the *MailOnline's* reliance on news agencies and all outlets' repeated use of XR creators as sources. The following chapters will provide greater depth to this analysis by examining the specific frames applied to XR in news and marketing.

References

Arik, M.B. and Çağlar, S. (2005) 'The Face of Consumption Society in the Press: Life Style Journalism', *Proceedings of the 3rd International Symposium Communication in the Millennium*, Chapel Hill, NC, 11–13 May. Available at: https://citeseerx. ist.psu.edu/viewdoc/download?doi=10.1.1.507.9260&rep=rep1&type=pdf (Accessed: 7 April 2021).

Bastos, M.T. (2019) 'Tabloid Journalism', in Vos, T.P., Hanusch, F., Geertsema-Sligh, M., Sehl, A. and Dimitrakopoulou, D. (eds.) *The International Encyclopedia of Journalism Studies*. Malden, MA: Wiley. doi:10.1002/9781118841570.

Bednarek, M. and Caple, H. (2012) *News Discourse*. London: Continuum International Publishing Group.

Bell, A. (1991) *The Language of News Media*. Oxford: Blackwell.

Bennett, L. (2016) 'News invests in augmented reality start-up', *NewsMediaWorks*, 21 April. Available at: https://newsmediaworks.com.au/news-invests-in-augmented-reality-start-up/ (Accessed: 7 March 2020).

Bettig, R.V. and Hall, J.L. (2012) *Big Media, Big Money*. 2nd edn. Plymouth: Rowman & Littlefield.

Bird, S.E. (2009) 'Tabloid Newspapers', in Sterling, C.H. (ed.) *Encyclopedia of Journalism*. London: SAGE, pp. 1361–1364.

Brennen, S.J., Howard, P.N. and Nielsen, R.K. (2020) 'Balancing Product Reviews, Traffic Targets, and Industry Criticism: UK Technology Journalism in Practice', *Journalism Practice*, [Preprint], pp. 1–18. doi:10.1080/17512786.2020.1783567.

Bull, A. (2010) *Multimedia Journalism: A Practical Guide*. Oxon: Routledge.

Caple, H. and Bednarek, M. (2013) *Delving into the Discourse: Approaches to News Values in Journalism Studies and Beyond*. Available at: https://reutersinstitute.politics.ox.ac.uk/sites/default/files/2018-01/Delving%20into%20the%20Discourse.pdf (Accessed: 16 October 2020).

Carvalho, A. and Burgess, J. (2005) 'Cultural Circuits of Climate Change in U.K. Broadsheet Newspapers, 1985–2003', *Risk Analysis*, 25(6), pp. 1457–1469. doi:10.1111/j.1539-6924.2005.00692.x.

Cogan, B. (2005) '"Framing Usefulness:" An Examination of Journalistic Coverage of the Personal Computer from 1982–1984', *Southern Journal of Communication*, 70(3), pp. 248–295. doi:10.1080/10417940509373330.

Cohen, S. (2002) *Folk Devils and Moral Panics*. 3rd edn. Oxon: Routledge.

Conill, R.F. (2016) 'Camouflaging Church as State', *Journalism Studies*, 17(7), pp. 904–914. doi:10.1080/1461670X.2016.1165138.

Critcher, C. (2003) *Moral Panics and the Media*. Buckingham: Open University Press.

Davies, N. (2009) *Flat Earth News*. London: Vintage.

Dimopoulos, K. and Koulaidis, V. (2002) 'The Socio-Epistemic Constitution of Science and Technology in the Greek Press: An Analysis of its Presentation', *Public Understanding of Science*, 11, pp. 225–241.

Entman, R.M. (1991) 'Framing U.S. Coverage of International News: Contrasts in Narratives of the KAL and Iran Air Incidents', *Journal of Communication*, 41(4), pp. 6–27.

Erjavec, K. (2004) 'Beyond Advertising and Journalism: Hybrid Promotional News Discourse', *Discourse & Society*, 15(5), pp. 553–578.

Fairclough, N. (1989) *Language and Power*. Harlow: Longman Group UK Limited.

Fairclough, N. (1993) 'Critical Discourse Analysis and the Marketization of Public Discourse: The Universities', *Discourse & Society*, 4(2), pp. 133–168.

Forde, S. and Johnston, J. (2013) 'The News Triumvirate: Public Relations, Wire Agencies and Online Copy', *Journalism Studies*, 14(1), pp. 113–129. doi:10.1080/1461670X.2012.679859.

Fuchs, P., Guez, J., Hugues, O., Jégo, J., Kemeny, A. and Mestre, D. (2017) *Virtual Reality Headsets — A Theoretical and Pragmatic Approach*. London: CRC Press.

Galtung, J. and Ruge, M.H. (1965) 'The Structure of Foreign News: The Presentation of the Congo, Cuba and Cyprus Crises in Four Norwegian Newspapers', *Journal of Peace Research*, 2(1), pp. 64–90. doi:10.1177/002234336500200104.

Gamson, W.A. (2001) 'Foreword', in Reese, S.D., Gandy, O.H. and Grant, A.E. (eds.) *Framing Public Life*. London: Lawrence Erlbaum Associates, pp. ix–si.

Gans, H.J. (1980) *Deciding What's News*. London: Constable.

Goggin, G. (2006) *Cell Phone Culture: Mobile Technology in Everyday Life*. Oxon: Routledge.

Hall, S., Critcher, C., Jefferson, T., Clarke, J. and Roberts, B. (1978) *Policing the Crisis*. London: Macmillan.

Hallahan, K. (1999) 'Seven Models of Framing: Implications for Public Relations', *Journal of Public Relations Research*, 11(3), pp. 205–242. doi:10.1207/s1532754xjprr1103_02.

Hanusch, F. (2012) 'Broadening the Focus: The Case for Lifestyle Journalism as a Field of Scholarly Inquiry', *Journalism Practice*, 6(1), pp. 2–11. doi:10.1080/17512786.2011.622895.

Harcup, T. and O'Neill, D. (2001) 'What Is News? Galtung and Ruge Revisited', *Journalism Studies*, 2(2), pp. 261–280. doi:10.1080/14616700118449.

Harcup, T. and O'Neill, D. (2017) 'What Is News? News Values Revisited (Again)', *Journalism Studies*, 18(12), pp. 1470–1488. doi:10.1080/1461670X.2016.1150193.

Harro-Loit, H. and Saks, K. (2006) 'The Diminishing Border between Advertising and Journalism in Estonia', *Journalism Studies*, 7(2), pp. 312–322. doi:10.1080/14616700500533635.

Kelly, J.P. (2009) 'Not so Revolutionary after All: The Role of Reinforcing Frames in US Magazine Discourse about Microcomputers', *New Media & Society*, 11(1–2), pp. 31–52. doi:10.1177/1461444808100159.

Kim, H., Chan, H.C. and Gupta, S. (2007) 'Value-based Adoption of Mobile Internet: An Empirical Investigation', *Decision Support Systems*, 43(1), pp. 111–126. doi:10.1016/j.dss.2005.05.009.

Kovach, B. and Rosenstiel, T. (2014) *The Elements of Journalism*. 3rd edn. New York: Three Rivers Press.

Lewis, J., Williams, A. and Franklin, B. (2008) 'A Compromised Fourth Estate?', *Journalism Studies*, 9(1), pp. 1–20. doi:10.1080/14616700701767974.

Lewis, J., Williams, A., Franklin, B., Thomas, J. and Mosdell, N. (2008) *The Quality and Independence of British Journalism*. Available at: https://orca.cf.ac.uk/18439/1/Quality%20%26%20Independence%20of%20British%20Journalism.pdf (Accessed: 16 March 2018).

Lipworth, W., Kerridge, I., Morrell, B., Forsyth, R. and Jordens, C.F.C. (2015) 'Views of Health Journalists, Industry Employees and News Consumers About Disclosure and Regulation of Industry-Journalist Relationships: An Empirical Ethical Study', *Journal of Medical Ethics*, 41, pp. 252–257.

Markey, P.M. and Ferguson, C.J. (2017) 'Teaching Us to Fear', *American Journal of Play*, 10(1), pp. 99–115.

Marwick, A.E. (2008) 'To Catch a Predator? The MySpace Moral Panic', *First Monday*, 13(6). doi:10.5210/fm.v13i6.2152.

McKernan, B. (2013) 'The Morality of Play: Video Game Coverage in *The New York Times* from 1980 to 2010', *Games and Culture*, 8(5), pp. 307–329. doi:10.1177/1555412013493133.

McManus, J.H. (1995) 'A Market-Based Model of News Production', *Communication Theory*, 5(4), pp. 301–338.

McManus, J.H. (2009) 'The Commercialization of News', in Wahl-Jorgensen, K. and Hanitzsch, T. (eds.) *The Handbook of Journalism Studies*. Oxon: Routledge, pp. 218–233.

News Corp (2017) *PropTiger and Housing.com Come together to Become India's Largest Digital Real Estate Company* [Press Release]. 10 January. Available at: https://newscorp.com/2017/01/09/proptiger-and-housing-com-come-together-to-become-indias-largest-digital-real-estate-company/ (Accessed: 7 March 2020).

Pan, Z. and Kosicki, G.M. (1993) 'Framing Analysis: An Approach to News Discourse', *Political Communication*, 10, pp. 55–75.

Richardson, J.E. (2007) *Analysing Newspapers*. Basingstoke: Palgrave.

Rogers, E.M. (2003) *Diffusion of Innovations*. 5th edn. New York: Free Press.

Rogers, R. (2013) 'Critical Essay — Old Games, Same Concerns', *Technoculture*, 3. Available at: https://tcjournal.org/drupal/vol3/rogers (Accessed: 10 October 2016).

Rosenkranz, T. (2016) 'Becoming Entrepreneurial: Crisis, Ethics and Marketization in the Field of Travel Journalism', *Poetics*, 54, pp. 54–65. doi:10.1016/j.poetic.2015.09.003.

Shoemaker, P.J. and Reese, S.D. (2014) *Mediating the Message in the 21st Century*. Oxon: Routledge.

Steinicke, F. (2016) *Being Really Virtual*. Cham, Switzerland: Springer.

Tankard, J.W. (2001) 'The Empirical Approach to the Study of Media Framing', in Reese, S.D., Gandy, O.H. and Grant, A.E. (eds.) *Framing Public Life*. Oxon: Routledge, pp. 95–105.

van Dijk, T.A. (1988) *News as Discourse*. New Jersey: Lawrence Erlbaum Associates.

Williams, A. and Clifford, S. (2009) *Mapping the Field: Specialist Science News Journalism in the UK National Media*. The Risk, Science and the Media Research Group, Cardiff University. Available at: http://orca.cf.ac.uk/18447/ (Accessed: 8 April 2021).

Witschge, T., Fenton, N. and Freedman, D. (2010) *Protecting the News: Civil Society and the Media. Carnegie UK Trust*. Available at: www.carnegieuktrust.org.uk/publications/protecting-the-news-civil-society-and-the-media/ (Accessed: 5 March 2018).

Wojdynski, B.W. (2016) 'Native Advertising: Engagement, Deception, and Implications for Theory', in Brown, R., Jones, V.K. and Wang, B.M. (eds.) *The New Advertising: Branding, Content and Consumer Relationships in a Data-Driven Social Media Era*. Santa Barbara, CA: Praeger/ABC Clio, pp. 203–236.

Zelizer, B. and Allan, S. (2010) *Keywords in News and Journalism Studies*. Berkshire: Open University Press.

4 Framing the Characteristics of XR

Immersion and Transcendence

Having explored the overall characteristics of XR news, this is the first of four chapters that combines quantitative and qualitative data to further analyse how XR has been framed in both the news and marketing. It first provides some more methodological details about how frames were identified in the study, followed by introducing some information about the first frame category: Conceptualisation. The chapter then moves on to an in-depth analysis of the two frames that were identified in this category in relation to XR: Immersive and Transcendent. The theoretical literature introduced in Chapter 1 (frame-building, the hierarchy of influences model and diffusion of innovations theory) is utilised to critically discuss the results.

Approaches to identify frames can be either deductive or inductive. A deductive approach involves analysing texts to look for specific frames that are defined before the research begins (Matthes and Kohring, 2008). The drawback to this is that it risks missing important frames that may be present within the texts, particularly for evolving issues such as emerging technologies. Instead, an inductive approach to identifying frames is much more suitable in this area as this allows frames to emerge throughout the course of analysis (de Vreese, 2005). Inductive approaches to framing analysis have been criticised for lacking replicability and comparability (e.g. Tankard, 2001). However, the utilisation of the frame categories presented in this book allows researchers to avoid both the prescriptive nature of deductive frame identification and the issues around comparability. That is to say, it is possible to identify new and emerging frames while still being able to make comparisons between the set categories.

In order to identify frames inductively, this study carried out a three-step process. The first step involved highlighting themes (rather than frames) in the texts. An in-depth examination of each news article and marketing material was carried out, highlighting themes throughout. This resulted in a list of 110 themes that appeared in the news articles and 68 that appeared in the marketing materials. The second step was to organise these specific themes into more easily manageable groups. This involved synthesising any related themes across both samples into broader themes. For instance, the themes of

DOI: 10.4324/9781003375814-4

Table 4.1 Four categories of frames appearing in XR discourse

(1) Frames conceptualising XR	(2) Newness frames	(3) User experience frames	(4) Evaluative frames
Immersive	Different and unique	Social	Important
Transcendent	Revolutionary and transformative	Easy to use	Successful
	Advanced and high-quality	Comfortable	Affordable
			Much-anticipated

"intuitive", "convenient", "unobtrusive" and "natural" all broadly referred to the ease of using XR devices. Therefore, they were grouped together under the theme of "ease to use". Any themes that did not relate to a broader category remained separate.

Finally, the third step involved revisiting these themes to see which of them could be defined as frames. This study treated a frame as more than a theme in that framing involves salience. That is, it involves "making a piece of information more noticeable, meaningful, or memorable to audiences" (Entman, 1993: 53). Once the themes had been identified in the previous step, they were revisited to determine whether they could be considered frames based on this idea of salience. On the one hand, this meant examining how many times these themes appeared in the news and marketing, since themes that only appeared a handful of times would not be particularly salient. Additionally, this also involved examining all instances of a theme for any framing devices that might have been used. As suggested by Linström and Marais (2012), framing devices can be rhetorical (issues of language, such as word choice, metaphors and exemplars) or technical (the elements comprising a text, such as how it is categorised, headlines, photographs, layout and the use of quotes). If a theme was made particularly salient by repetition and through such framing devices, it was considered a frame. This resulted in the identification of 12 frames that were categorised into four frame categories, as shown in Table 4.1.

The first category within this framework is Conceptualisation. As an actor in the social construction of technology, the news media play an important role in how an innovation is conceptualised (McKernan, 2013), highlighting the importance of studying these types of frames. This category includes frames that are related to concepts specific to the technology under study. Since "[i]nnovation is about change" (Krumsvik et al., 2019: 193), each emerging technology has its own features that differentiate it from existing products. It is possible for there to be similarities between different technologies, but taking an inductive, grounded approach to identifying frames is key to ensure that researchers do not miss new frames as a result of being restricted to pre-defined frames.

When analysing media texts about technology, researchers can identify frames within the Conceptualisation category by asking questions such as:

- What specific characteristics of the technology have been emphasised within the discourse?
- What are portrayed as the key features of the technology?
- What concepts in the academic literature are associated with the technology and does this align with its media representations?

The current study found that two frames worked towards conceptualising XR: Immersive and Transcendent. The rest of this chapter discusses the framing devices used to construct the Immersive and Transcendent frames in XR news and compares this with XR promotional materials.

Immersive

Chapter 2 introduced immersion and presence as the two key features of XR technology. If effective, XR experiences provide users with a sense of immersion and presence (Evans, 2019). This allows them to believe they are inside a virtual environment, thus resulting in the user trying to interact with it as such (Lombard and Ditton, 1997; Steptoe, Julier and Steed, 2014). The framing analysis uncovered that an Immersive frame was applied to XR, representing the technology as able to make the user feel a sense of immersion and presence when experiencing XR.

Word choices and keywords can act as framing devices (Entman, 1993; Linström and Marais, 2012) to construct a frame. Therefore, the frequency of words pertaining to each frame were recorded to give some indication as to the prevalence of these frames. Examining this data shows that words relating to the Immersive frame were mentioned the most out of any frame-based category (see Appendix 5). Terms in the "immersive" category appeared 1,457 times in 56 percent of articles. Aside from this, words in the "advanced and high-quality" category were mentioned in the second largest portion of articles, though substantially less than words in the "immersive" category (30 percent). This demonstrates just how prominent the Immersive frame was. Furthermore, examining the use of specific terms provides additional insight into the prevalence of the Immersive frame. Out of all individual search terms (across all categories), the stem *immers** (e.g. immersion, immersive) was used, by far, the most times (963) and in the most articles (45 percent). For comparison, the second most used term, *excit** (e.g. exciting, excitement) appeared 246 times in 18 percent of articles – significantly less than *immers**, though this will be discussed in Chapter 7. Thus, these figures indicate that it was very common for articles to apply the Immersive frame to XR.

Moreover, all news outlets used words in the "immersive" category more than any other frame category. Likewise, every news outlet used *immers** more than any other search term. This shows that the media organisation factor of the hierarchy of influences (Shoemaker and Reese, 2014) has not had much impact on the prevalence of the Immersive frame, since there is little difference in how often the three news outlets presented XR as Immersive. On the topic of innovation news, Nordfors states that "[m]any who read a news item feel that new knowledge is confirmed when others discuss it or when they see it again in a different news outlet. Such news is more likely to be accepted as fact" (2009: 21). Therefore, the reiteration of the Immersive frame in multiple news outlets increases the likelihood that readers will come to accept this framing of XR.

On the other hand, the use of terms in the "immersive" category varied quite dramatically between articles focusing on VR and those focusing on AR/MR. Words in the "immersive" category appeared in 63 percent of VR articles and 23 percent of AR/MR articles. While 23 percent is still a substantial amount, it is clear that the Immersive frame was used more often in relation to VR products than AR/MR products. This finding also helps to explain the variation in how often these terms appeared across the sample period. The use of words in the "immersive" category increased dramatically in 2014 and continued to appear in at least half of articles for the rest of the sample period (see Appendix 6). The shift in 2014 coincides with the year the news articles started to focus more on VR products than AR/MR, which were most common in the first two years (see Chapter 3). Since VR works by replacing the user's view of the real world with a virtual environment (Brigham, 2017), the concepts of immersion and presence are more important in VR experiences than they are for AR/MR products. Therefore, the technological characteristics of the devices appear to have impacted the way they are framed in the news coverage.

As well as word choice, additional rhetorical framing devices were used to construct the Immersive frame. For instance, examining the stem *immers** within the context of the news articles revealed that journalists used certain modifiers to emphasise the effectiveness of this immersion. One *MailOnline* article described a PlayStation VR demo as "incredibly immersive" (Spettigue, 2015). Additionally, journalists writing in the *Guardian* and the *MailOnline* both claim they were "completely immersed" (Stuart, 2015; Liberatore, 2016) during their HoloLens experiences. Here, the modifiers "incredibly" and "completely" work to emphasise the Immersive frame by insinuating that the sense of immersion is of a high quality. Significantly, a similar technique was observed in the marketing of Oculus Rift which described the product as "truly immersive" in its Kickstarter campaign (Kickstarter, 2012) and the press release about that campaign (Oculus VR, 2012), implying it is a highly immersive experience.

A sense of immersion and presence is not guaranteed by all XR products and experiences, but instead requires combining "a number of elements of different sensory stimuli and preparedness on the part of the VR user" (Evans, 2019: 50). Despite this, the news and marketing discourses have both claimed the technology allows users to be highly immersed in a virtual environment. Since the news is the general public's main source of information about emerging technologies (Whitton and Maclure, 2015; Williams, 2003), this reinforces the message present in the marketing and thus supports the promotion of XR.

While *immers** and *excit** were the two most used stems in the news sample, it is also significant that the only other individual word used in at least 10 percent of articles was *transport**. This term appeared 187 times in 13 percent of articles. The use of this term was recorded when it implied that an XR user could be metaphorically transported with the technology, rather than referring to literal forms of transport. This, therefore, demonstrates the use of the transportation metaphor as a framing device to portray XR as Immersive. In more detail, a *Guardian* article about PlayStation VR stated: "All perception of your real-world surroundings become removed as you place the headset on and are instantly transported to another world" (Gibbs, 2014). Suggesting XR can transport a person to another world creates the impression that the user believes they are in that virtual world, highlighting the idea of immersion.

Importantly, regarding the relationship between the news and the marketing, it was found that this same metaphor was employed in the promotional materials of XR. For instance, both Oculus Rift and Gear VR marketing used the word "transport". On the Gear VR page of the Oculus website, it was stated: "It's easy to transport yourself with the Gear VR" (Oculus VR, 2015b). A post on the Oculus Rift Facebook page touted: "Transport yourself to Japan's spectacular Motenashi Dome of Kanazawa Station in 360 [degrees] on Rift and Gear VR" (Oculus, 2017c). Similarly, in a table detailing the differences between VR, AR and MR on the HoloLens website, VR and MR were said to be able to "transport you to a virtual world" (Microsoft, 2016a). Relatedly, a major part of Oculus Rift marketing alluded to the metaphor of transportation without using the word itself. The tagline of the product was "Step into the game", or sometimes "Step into the Rift", which appeared on their Kickstarter campaign (Kickstarter, 2012), as well as their website (Oculus VR, 2015a, 2015c) and promotional videos (Oculus, 2015a, 2017a). The idea of "stepping into" another world highlights the sense of being transported to another place. That is to say, users will feel as if they are really there – immersed in a virtual world. These findings highlight a similarity between XR news and marketing in the sense that both discourses employ metaphors as framing devices to highlight immersion and presence. According to Luokkanen, Huttunen and Hildén, metaphors have a particularly important role when communicating about topics that are "novel, abstract or without stabilized context" (2014: 967). Thus, as XR was in the early stages of diffusion during this period, the use of metaphors to highlight immersion could be particularly effective in emphasising the Immersive frame. This is enhanced further since the metaphors are repeated in both news and marketing discourses.

Within the news articles specifically, other descriptions of XR experiences framed the technology as Immersive through the use of active verbs. For example, a *MailOnline* article about PlayStation VR claimed the device allows "players to fly like an eagle, drive sports cars in high-speed races, and explore castles" (Plummer and O'Hare, 2016). The words "fly", "drive" and "explore" suggest users will feel as if they are actually carrying out these actions, thus implying immersion. A similar example can be observed in the following two paragraphs which appeared in a side-note about Oculus Rift in the *MailOnline*:

> While the resolution still doesn't give the feeling of quite being in the real world, it does make you think you are actually in a virtual world.
> During several demonstrations we entered a vast dungeon and flew through space.
>
> (Prigg, 2014a)

First, although the journalist argues that the experience does not appear realistic, it is still said to make the user believe that they are in a virtual world, framing it as Immersive. Additionally, instead of using a sentence such as "we saw a vast dungeon and a space scene", the journalist writes that they "entered" a dungeon and "flew" through space. Though the journalists did not physically perform these actions, using active verbs suggests the experiences felt real enough that they believed they were in these spaces.

Furthermore, this *MailOnline* side-note appeared not just in one article but in 16 different *MailOnline* reports from March 2014 to January 2016. Sheafer, Shenhav and Amsalem's frame repetition hypothesis claims that "the frames that have greater influence on public opinion are those that are repeated more frequently" (2018: 264). This idea is shared by Fredlin who states that "incessantly repeated" frames can "have considerable control over how people think about various issues and events" (2001: 272). Therefore, the reiteration of this side-note has put particular emphasis on the Immersive frame. Moreover, regarding the hierarchy of influences model (Shoemaker and Reese, 2014), the *MailOnline*'s routine practice to repeat its side-notes in multiple articles has impacted the frame-building process in this instance.

As mentioned in Chapter 2, immersion and presence go hand-in-hand with each other. However, the word *presence* was rarely used within the news articles (appearing in only 3 percent), perhaps to avoid the use of jargon. Alternatively, both immersion and presence were depicted using technical framing devices in the form of imagery. If a user feels a sense of presence, they tend to respond to the virtual environment in a realistic way (Lombard and Ditton, 1997; Steptoe, Julier and Steed, 2014). With this in mind, the news articles depicted presence by including pictures of users holding out their hands as if instinctively attempting to interact with the virtual world they are seeing through a VR headset. Examples of such images can be seen in the *Guardian* focusing on Google Cardboard (Shubber, 2014) and Oculus Rift (Johnston, 2015), as well as the *MailOnline*'s coverage of Oculus Rift (Woollaston, 2015). Importantly, considering the relationship between the news and marketing samples, the same type of imagery also appeared in XR promotional material (for example, on the website for Oculus Rift (Oculus VR, 2016) and social media posts about Gear VR on the Oculus Facebook page (Oculus, 2015b, 2016)). This shows that, not only have XR company owners been used as sources the most in the news articles (see Chapter 3), but similar imagery has been used to portray presence even when it does not originate from these company owners. This further reinforces the desired frames of those invested in the success of XR.

The same can be said for images depicting immersion – both samples used visuals to represent this concept. However, images representing immersion were not as visually similar in the two samples as they were in regards to presence. In the news articles, images of users wearing headsets often had very plain backgrounds with a large amount of empty space above or in front of the headset (e.g. Evans, 2016; Hern, 2016; Spettigue, 2015). What the user is viewing cannot be seen in the images, but a space is still left blank for this. The empty space indicates that the user is unaware of anything apart from what they are seeing through the headset, which ultimately implies they are immersed. On the other hand, in the marketing materials, immersion is depicted using more visually complex techniques. For example, in a video advertisement for Oculus Rift (Oculus, 2017a), four different users are shown experiencing VR. In each instance, after they put on the headset, their living room splits open and the virtual world can be seen to surround them. Again, this suggests that the XR experience is convincing enough that the user believes it is real. Thus, in both samples, visuals have been used as framing devices to construct the Immersive frame, though in quite different ways.

Previous research has suggested that images have a significant impact on viewers' attention and perceptions of a topic (Jenner, 2012; Müller, Kappas and Olk, 2012). Coleman agrees, stating it is thought that "the unique, vivid features of pictures make them more readily available in memory; thus, images exert a more powerful influence on memory and perceptions than text" (2010: 243). Therefore, the use of images as technical framing devices, in combination with the other techniques discussed above, highlights the strength of the Immersive frame in both the news and marketing samples.

Overall, these findings show that the Immersive frame was very prominent in XR news coverage and marketing materials, particularly in relation to VR. Nordfors states that all innovations come with "new words and stories" (2009: 18). Although immersion as a concept is not new, the technologically induced immersion offered by XR is different to what was previously possible (Ryan, 2015). By introducing and spreading this new language, journalism "speeds up the introduction of new things, enabling people to discuss them before they are widely spread. This facilitates introduction" (Nordfors,

2009: 18). Thus, the prominence of the Immersive frame supports the introduction of this concept and, thus, the diffusion of XR generally.

Moreover, while it is the aim of a VR experience to be immersive, a sense
of immersion is certainly not a given simply with the use of a VR headset
(Evans, 2019). Nevertheless, both the news articles and marketing materials
have presented XR as highly immersive. Since immersion is the key selling
point of VR (Evans, 2019), it is not surprising that this frame was present
in XR marketing. However, the existence and strength of this frame in XR
news coverage reinforces the idea that XR achieves its aim of creating an
immersive experience. In other words, the news articles effectively aid the
promotion of XR since they support its unique selling point. It appears that
the use of native advertising through links to retailers and information about
where to buy products as discussed in Chapter 2 is not the only way XR news
has been marketised (Fairclough, 1993). Again, there seems to be a blurring
of news and promotional content as was identified by Erjavec (2004), Lewis,
Williams and Franklin (2008) and Chyi and Lee (2018), amongst others.

Despite there being several concerns surrounding VR related to its immersive capabilities (such as cybersickness and physical isolation; see Chapter 2),
the Immersive frame did not draw on these issues and instead presented
immersion as something positive and enjoyable. In line with the results from
the previous chapter, this is further indication that the news coverage did not
foster a moral panic for XR during its introduction. Instead, the use of the
Immersive frame supports the interests of XR companies since it presents
the key selling point of the technology in a positive light. This is particularly
the case since the quantitative data suggests Immersive was the most commonly used frame in the news articles.

Transcendent

Another concept relevant to XR is transcendence, involving "going beyond"
(Anderson, 2003), in this case, what was possible without XR technology.
Chan states that "the hype and hope that are associated with transcending
the physical body for exalted wonderment in virtual realities can be traced
back to the seventeenth century" (2014: 8). Indeed, this study found that the
current generation of XR is no exception, with Transcendent emerging as a
frame in both the news and marketing discourses. This involved presenting
XR as able to overcome physical or bodily limitations, as well as limitations of
previous technologies. The current section examines the framing devices used
to construct this frame in the news articles and its relation to XR marketing.

To begin, quantitative data demonstrates the prominence of the
Transcendent frame in each news outlet. Although not used as often as
words in the "immersive" category, terms in the "transcendent" category
appeared in 15 percent of articles overall. There was little difference between
the news outlets in how often they used words in this group. Thus, as with
the Immersive frame, the prevalence of the Transcendent frame does not

appear to differ significantly per news outlet. This shows that, within the hierarchy of influences model (Shoemaker and Reese, 2014), the media organisation reporting on XR has not had much effect on the strength of the Transcendent frame.

Moreover, these terms appeared in a fairly consistent portion of articles every year, with the exception of a trough in 2013. On the other hand, there were substantial differences between VR and AR/MR articles. Out of the subset of VR articles, 13 percent included words from the "transcendent" category, whereas 25 percent of AR/MR articles used such terms. Therefore, the Transcendent frame appears to have been applied more so to AR/MR products than VR, although the frame broadly gained the most traction from 2014 onward. Since to "augment" is "to achieve a higher or intensified state of reality" (Ariel, 2017: 31), it is clear that the idea of transcendence is highly relevant to AR and, due to its close links, MR. These figures suggest that this is reflected in the news media and shows that the technological characteristics of these devices have impacted their framing.

Although words in the "transcendent" category appeared in 15 percent of articles overall, it is important to note that the stem *transcend** itself was not used in any articles. As with the lack of the term *presence*, this may be due to the typical journalistic practice to use lay terms that a wider audience would understand. Certainly, Carlson states that "the public relies on journalism to translate complex discourses into understandable ones" (Carlson, 2017: 43). Instead of using *transcend**, the idea of transcendence was explained in other ways, using simpler terms. For example, the stem *improv** (e.g. improve, improves) appeared in 5 percent of articles, *enhance** was used in 4 percent and *beyond* also appeared in 5 percent. Therefore, related to the hierarchy of influences model (Shoemaker and Reese, 2014), the routine practices of journalism have impacted the construction of the Transcendent frame by affecting the specific words chosen when referring to this concept.

In addition to word choices, exemplars were used as rhetorical framing devices when presenting XR as Transcendent. Specifically, XR's transcendental capabilities were related to a wide range of areas. For instance, the technology was said to be able to improve education (Bloxham, 2013), the wellbeing of the terminally ill (Block, 2014; Boyle, 2017), the lives of those with autism (Griffiths, 2014a) and the detection of breast cancer (Griffiths, 2014b). Furthermore, other articles claimed XR could enhance rollercoasters (Barnes, 2017), storytelling (AFP, 2015) and military safety and intelligence (Burrows, 2015; Prigg, 2014b), amongst other areas. Since improving or enhancing something means it has become better than it was previously, this links to overcoming limitations and thus transcendence. Shen states that how persuasive a frame is "will likely depend on how the messages interact with individuals' own predispositions or knowledge structure" (2004: 126). Thus, relating transcendence to this broad selection of areas increases the salience of the frame by making it relevant to a range of readers. In other words, some readers will see the value in improving the lives of those with autism and others will see the value of enhancing military intelligence. Mentioning a wide array of areas that XR will positively impact increases the portion of the audience this frame will resonate with, thereby increasing its strength.

Moreover, these exemplars were enhanced further with the use of a technical framing device: quotations from sources that are established in the technology industry. Electronics company LG was quoted in one article claiming that VR could "improve lives" (Griffiths, 2015) and Apple CEO Tim Cook highlighted the same point in relation to AR (Liberatore, 2017a). Go, Jung and Wu state that "the credibility of information is often determined by the believability of its source" (2014: 359). Due to their status in the technology industry, these sources could be seen as credible sources by readers. Therefore, journalists' decision to use such sources increases the strength of the Transcendent frame. Moreover, these sources can be seen as frame advocates, part of the social institutions factor of the hierarchy of influences (Shoemaker and Reese, 2014). Frame advocates use frames strategically with the aim of getting their preferred frame into the news (Van Gorp, 2010), in order to achieve a certain outcome (Kee, Hassan and Ahmad, 2012). These examples show that an XR company owner (LG) and a technology specialist (Tim Cook) have acted as frame advocates for the Transcendent frame, demonstrating their impact on the frame-building process.

Considering the relationship between XR news and promotional content, it was found that the marketing used similar exemplars to depict the Transcendent frame. The HoloLens website claimed the device can enable "you to make decisions more confidently [and] work more effectively" (Microsoft, 2015a), particularly in product design. Moreover, a later version of the website claimed the device can "increase students' engagement and understanding of abstract concepts" (Microsoft, 2017) in relation to education. Moreover, when Google Glass was relaunched as an enterprise product in 2017, its new website highlighted several ways it could be used to improve productivity and efficiency. For instance, a quote on the website from one business that had used Google Glass stated: "Glass really gives our operators the ability to do their jobs faster, smarter, and safer" (Google, 2017). The repetition of comparative words (faster, smarter and safer) combined with "really" stresses the transcendent effect of Google Glass, thus increasing the power of the exemplar in highlighting the Transcendent frame. This shows that both samples framed XR as Transcendent and used exemplars to do so. Van Gorp argues that the more frames "are confirmed by further information, or by congruent framing devices, the more difficult it becomes to refute or change them by counterframing" (2007: 69). Therefore, the appearance of the Transcendent frame in the news discourse reinforces the frame in the marketing and vice versa, showing that the news supports XR marketing.

Additional similarities were also uncovered between the news and marketing samples. One rhetorical framing device used was the argument that XR transcends the limitations of the traditional screen interface. For instance, Magic Leap was described in the *Guardian* as aiming to get "rid of dependence on screens" (Cumming, 2014). Similarly, AR contact lenses being developed were said to have the potential to "do away with TV screens" in the headline of a *MailOnline* article (Zolfagharifard, 2014). Another report in the *Guardian* about the HoloLens Minecraft application quoted Microsoft's corporate vice-president Kudo Tsunoda, stating "there's a level of immediacy and intimacy that goes beyond anything you can experience while sitting in front of a television screen" (Stuart, 2015). In these excerpts, the Transcendent frame is created by relating the widely understood concept of the traditional screen to XR – claiming XR will improve upon the older technology. Furthermore, in the last case, the exemplar originates from a source that is established in the field (Kudo Tsunoda). This gives the statement credibility (Go, Jung and Wu, 2014) and further emphasises the frame itself.

In XR marketing, the same framing device was used. For example, on Magic Leap's website homepage, the interface of the device was described as allowing users to "break free from outdated conventions of point and click interfaces, delivering a more natural and intuitive way to interact with technology" (Magic Leap, 2017). "Breaking free" implies that the current way of interacting with technology is restrictive and limiting, whereas MR (and Magic Leap) can transcend this. Similarly, in an early promotional video for HoloLens, a creator of the device stated that the way we usually interact with technology (through a screen) is a very "cold" and limited experience (Microsoft, 2015b). She continued to say that they aim to overcome this with HoloLens, which goes "beyond the screen". Indeed, "go beyond the screen" was a very common phrase in HoloLens marketing, appearing in various promotional materials (including videos and across the website). It is therefore significant that the stem *beyond** was used in 5 percent of articles, as mentioned above. The idea of going beyond something that already exists to a superior experience appears to be shared between the news and the marketing of XR. Importantly, framing XR as Transcendent in this way highlights its relative advantage in comparison to other technologies. As one of the five perceived attributes of innovations, Rogers defines relative advantage as "the degree to which an innovation is perceived as better than the idea it supersedes [...] The greater the perceived relative advantage of an innovation, the more rapid its rate of adoption will be" (2003: 15). With this in mind, framing XR as Transcendent could potentially support its diffusion, particularly since this is reinforced by both the news and marketing discourses.

As well as transcending previous technology, another rhetorical framing device involved claiming that XR can be used to do things that would be difficult or even impossible without the technology. For example, a *Guardian* article claimed that:

One of the key selling points for VR technology is its ability to put you in places you're unlikely to visit in the flesh, whether too expensive, too dangerous, out of bounds because of mobility issues or just because you don't like flying.

(Dredge, 2016)

Here, focusing on travel, VR users are said to be able to experience locations virtually that they may be unable to physically due to various factors. Regarding other uses, one *MailOnline* article described a demonstration of the Fove headset in which "a bed-ridden grandmother wears a Fove headset to 'attend' her grandson's wedding" (Associated Press, 2016). In the same article, it was noted that by using the headset, "a young man with spinal muscular atrophy, an illness that has weakened his arms and fingers, used eye movements to play a piano". Across these examples, XR is framed as allowing users to transcend certain limitations, whether that is money, risk, mobility issues or physical disability. This, again, demonstrates the use of exemplars as framing devices in the construction of the Transcendent frame.

Moreover, the same framing device was also present in XR marketing, though to a more extreme level. Aside from Google Glass, the promotional materials of every device analysed highlighted the Transcendent frame by implying that the products allow the impossible to become possible. Firstly, the HoloLens website explained that the headset allows NASA scientists to: "work as if they can walk on the surface of Mars, an experience previously impossible" (Microsoft, 2016b). Another sentence was used regarding car manufacturer Volvo in which HoloLens was said to bring "its cutting edge car features to life in ways never before possible" (Microsoft, 2016b). For the Magic Leap device, a post on their Facebook page included a quote from one of the developers with a similar sentiment: "Mixed reality is the mixture of the real world and virtual worlds so that one understands the other. This creates experiences that cannot possibly happen anywhere else" (Magic Leap, 2016). Moving on to VR, the Oculus Facebook page mentioned that their VR For Good initiative "explores VR's ability to inspire and make people feel things they might have previously thought were impossible" (Oculus, 2017b). Each of these excerpts present XR as allowing certain experiences to be had that are "impossible" without the technology, thus framing XR as Transcendent.

Furthermore, the idea of the impossible becoming possible was perhaps the strongest in a Gear VR video advert that featured an ostrich using the device to learn how to fly. In the advert, when the ostrich puts on the headset, it is shown a flight simulator. Since ostriches are known to be one of the few birds that cannot fly, this allows the ostrich to experience something it could not in reality. As the video continues, the ostrich attempts to fly in the real world while using the headset, without any success. At the end of the advert, the ostrich is shown without the headset, finally taking off into the sky in the real world. Thus, the advert insinuates that the ostrich is able to accomplish something that was previously thought impossible, with the help of VR. This is emphasised by the text that appears at the end of the video: "We make what can't be made", followed by "So you can do what can't be done" and finally the hashtag #DoWhatYouCant. These words extend the transcendent effect from allowing an ostrich to fly to suggesting that Gear VR can allow anyone to do anything they could not previously. It is clear that Samsung intended to present transcendence as one of the key features of its device, making it even more significant that this frame also appeared in the news articles.

Indeed, it was even found that a news article from the *MailOnline* was dedicated to writing about this Gear VR advert (Liberatore, 2017b). The *MailOnline*'s interpretation of this advert provides valuable insight into how XR marketing has been treated in the news, particularly relating to transcendence. The report opened with the following paragraphs:

Samsung has given an ostrich the ability to fly with the power of virtual reality.

The South Korean firm has released a new commercial that highlights an ostrich strapping on a headset playing a flight simulation – **giving the large bird the courage to spread its wings and take to the sky.**

Ending the clip with **#DoWhatYouCan't**, the advert for the Gear VR headset is a bid to convince viewers that **all of their dreams can come true in a virtual world** – they just need to purchase the technology.

(Liberatore, 2017b)

The sections in bold emphasise the rhetorical framing devices used by the journalist to create the Transcendent frame. To expand, the first sentence states that the ostrich has been able to do something that the species is known to be unable to do – fly – by using VR. This is reiterated in the second paragraph, with more detail. Calling the ostrich a "large bird" implies that this is a substantial feat due to its size and weight potentially making it more difficult to fly than other birds. Moreover, the third paragraph notes Samsung's hashtag that insinuates transcendence, as discussed above. The journalist summarises that Samsung is aiming to convince viewers that Gear VR would allow any user's dreams to come true. While dreams coming true does not necessarily relate to transcendence, this implies that it was the ostrich's dream to do something previously impossible, suggesting the same can apply to the dreams of real users. Therefore, highlighting this reinforces the message, and frame, from the advert itself.

After these introductory paragraphs, the article continued to describe the whole advert in depth. It then included several paragraphs with details about Gear VR itself, such as its specifications. This is an important finding because it potentially reduces the number of steps a consumer might go through when deciding whether to purchase a product. To expand, while there are several stages to the innovation decision-making process (Rogers, 2003), the news article has allowed the consumer to view the advert, read the journalist's interpretation of it and see information about the product all in one location. This condenses the decision-making process.

Additionally, featuring the advert in this news article increases its reach, allowing it to be seen by a wider audience of news readers rather than those who may see it on TV or online. As Schudson states: "when the media offer the public an item of news, [...] they not only distribute the report of an event or announcement to a large group, they amplify it" (1995: 19, quoted in Fuglsang, 2001: 197). Therefore, this is also significant because it shows that the news articles have not only reinforced the Transcendent frame present in XR marketing, but this article has done so by spreading the reach of the video advert itself. Still, only one article in the sample was entirely dedicated to discussing the marketing of an XR product. However, the findings discussed in Chapter 2 show that significant portions of multimedia (e.g. images and videos) originated from XR devices and applications. Thus, the Gear VR advert is not the only example of these companies having their promotional content included in the news, even if this was the one instance in which an entire article was dedicated to a marketing text. This perhaps suggests a blurring of news and promotional content, as was found by Erjavec (2004), Lewis, Williams and Franklin (2008), Chyi and Lee (2018) and others.

The existence of the Transcendent frame in both the news and marketing samples coincides with fictional portrayals of VR which have been found to highlight the transcendent capabilities of the technology (Chan, 2014; Taylor, 1997). However, it extends this past VR alone to the broader spectrum of XR. In addition, this news and marketing has not highlighted XR transcendence to the extreme extent of transhumanism as it appears in fiction. Rather, it is focused on how XR can improve or enhance the lives of its users, to make the impossible possible, or to transcend the traditional screen. In this way, it links to the "better than life" discourse found in the *T3* article that initially inspired this research. More broadly, the Transcendent frame also relates to what Roderick (2016) calls the discourse of technological satisfaction. This involves a focus on what a technological innovation "will do immediately to improve everyday life by overcoming some problem or limitation" (Roderick, 2016: 190). The Transcendent frame does this by representing XR as offering users the chance to do something that could not be done before, or at least to improve the quality of an experience compared to what was previously possible.

As such, the Transcendent frame presents XR positively, again showing that the frames used in the news articles to conceptualise XR do not contribute to building a moral panic about the technology. On the other hand, the Transcendent frame also appeared in the marketing materials, suggesting that it would be beneficial to XR companies to expand the reach of this frame to news coverage. Coupled with this, the salience of the Transcendent frame in the news increases the perceived relative advantage (Rogers, 2003) of XR, thus promoting its diffusion. Therefore, including the Transcendent frame in the news supports the commercial interests of XR companies by reinforcing the messages from the marketing and presenting a favourable view of the technology that should positively impact XR diffusion. In essence, the news acts as a promotional tool by presenting the technology this way.

Promotional News

The separation of news and promotional content is necessary to maintain the independence of the press (Lewis, Williams and Franklin, 2008). As opposed to traditional news, lifestyle journalism is sometimes seen as an extension of marketing (English and Fleischman, 2019; Kristensen, Hellman and Riegert, 2019). These results suggest that, despite being presented as news, XR articles share characteristics with lifestyle journalism. While reading content labelled as "news", readers are being subject to promotional frames, thus making them more susceptible to the messages in the same way that native advertising does. Both of these frames present XR positively, avoid critical considerations of XR (and, indeed, any moral panic style coverage) and reinforce the marketing efforts of XR companies.

Traditionally and idealistically, journalism has been seen as a fourth estate in which they "keep a skeptical eye on the powerful, guarding the public interest and protecting it from incompetence, corruption, and misinformation" (Norris and Odugbemi, 2010: 16). Within the fourth estate model, journalists take on a watchdog role (Hampton, 2010). In other words, they are responsible for holding those in power to account (Hansen, 2018), whether that may be governments, corporations or others. As Deuze argues, journalism should "share a sense of 'doing it for the public', of working as some kind of representative watchdog of the status quo in the name of people" (2005: 447). Indeed, Göpfert states that "the most important social task of journalism [is] to critically inform the public and act as a controlling entity" (2007: 224). Fourth estate journalism prioritises the interests of the general public by "provid[ing] information to help us understand the world and our position in it" (Richardson, 2007: 83). Fjæstad makes a similar point, claiming that journalists' "mission in Western societies is to serve their audiences, the citizens, by informing them about recent developments ('news'), and by naming and warning of insufficiencies of various kinds" (2007: 126). However, this certainly does not seem to have been applied in news of XR. Instead of prioritising the interests of the general public, the use of these frames supports the commercial interests of XR companies. Indeed, as the rest of this book will show, these are not the only frames that were shared between the news and marketing discourses.

References

AFP (2015) 'Virtual reality app brings crisis zones closer to home', *MailOnline*, 23 November. Available at: www.dailymail.co.uk/wires/afp/article-3329938/Virtual-reality-app-brings-crisis-zones-closer-home.html (Accessed: 13 February 2019).

Anderson, W.T. (2003) 'Augmentation, Symbiosis, Transcendence: Technology and the Future(s) of Human Identity', *Futures*, 35(5), pp. 535–546. doi:10.1016/S0016-3287(02)00097-6.

Ariel, G. (2017) *Augmenting Alice*. Amsterdam: BIS Publishers.

Associated Press (2016) 'The end of the joystick? VR headset with built in eye tracking 'adds empathy' to virtual worlds', *MailOnline*, 17 March. Available at: www.dailymail.co.uk/sciencetech/article-3496385/Startup-makes-virtual-reality-intuitive-eye-tracking.html (Accessed: 12 February 2019).

Barnes, L. (2017) 'Two trains smash into each other on rollercoaster at one of Spain's oldest theme parks injuring 26', *MailOnline*, 16 July. Available at: www.dailymail.co.uk/news/article-4701416/Two-trains-smash-rollercoaster.html (Accessed: 7 February 2019).

Block, J. (2014) 'Virtual reality headset allows terminally ill grandmother with cancer to experience the outdoors', *MailOnline*, 19 April. Available at: www.dailymail.co.uk/news/article-2608188/Virtual-reality-headset-allows-terminally-ill-grandmother-cancer-experience-outdoors.html (Accessed: 14 February 2019).

Bloxham, J. (2013) 'Augmented reality in education: teaching tool or passing trend?', *The Guardian*, 11 February. Available at: www.theguardian.com/higher-education-network/blog/2013/feb/11/augmented-reality-teaching-tool-trend (Accessed: 1 March 2019).

Boyle, S. (2017) 'Hospice helps bedridden terminally ill patients take a virtual walk through a park using headsets', *MailOnline*, 18 April. Available at: www.dailymail.co.uk/health/article-4420022/Hospice-helps-terminal-patients-virtual-technology.html (Accessed: 8 February 2019).

Brigham, T.J. (2017) 'Reality Check: Basics of Augmented, Virtual, and Mixed Reality', *Medical Reference Services Quarterly*, 36(2), pp. 171–178. doi:10.1080/02763869.2017.1293987.

Burrows, T. (2015) 'Generals will be able to direct battles using new Minority Report-style technology including 3D goggles and even virtual reality contact lenses', *MailOnline*, 10 May. Available at: www.dailymail.co.uk/news/article-3075485/Headsets-aid-military-commanders.html (Accessed: 13 February 2019).

Carlson, M. (2017) *Journalistic Authority: Legitimating News in the Digital Era*. New York: Columbia University Press.

Chan, M. (2014) *Virtual Reality: Representations in Contemporary Media*. London: Bloomsbury.

Chyi, H.I. and Lee, A.M. (2018) 'Commercialization of Technology News', *Journalism Practice*, 12(5), pp. 585–604. doi:10.1080/17512786.2017.1333447.

Coleman, R. (2010) 'Framing the Pictures in Our Heads: Exploring the Framing and Agenda-Setting Effects of Visual Images', in D'Angelo, P. and Kuypers, J.A. (eds.) *Doing News Framing Analysis*. Oxon: Routledge, pp. 233–261.

Cumming, E. (2014) 'Magic Leap: startup promises a leap forward for virtual reality', *The Guardian*, 25 October. Available at: www.theguardian.com/technology/2014/oct/25/virtual-reality-leap-forward-google (Accessed: 14 December 2018).

de Vreese, C. H. (2005) 'News Framing: Theory and Typology', *Information Design Journal + Document Design*, 13(1), pp. 51–62.

Deuze, M. (2005) 'What Is Journalism? Professional Identity and Ideology of Journalists Reconsidered', *Journalism*, 6(4), pp. 442–464. doi:10.1177/1464884905056815.

Dredge, S. (2016) 'The complete guide to virtual reality — everything you need to get started', *The Guardian*, 10 November. Available at: www.theguardian.com/tec hnology/2016/nov/10/virtual-reality-guide-headsets-apps-games-vr (Accessed: 20 December 2018).

English, P. and Fleischman, D. (2019) 'Food for Thought in Restaurant Reviews', *Journalism Practice*, 13(1), pp. 90–104. doi:10.1080/17512786.2017.1397530.

Entman, R.M. (1993) 'Framing: Towards Clarification of a Fractured Paradigm', *Journal of Communication*, 43(4), pp. 51–58.

Erjavec, K. (2004) 'Beyond Advertising and Journalism: Hybrid Promotional News Discourse', *Discourse & Society*, 15(5), pp. 553–578.

Evans, C.L. (2016) 'Virtual reality may look cool, but it will feel empty without community', *The Guardian*, 23 August. Available at: www.theguardian.com/commentisfree/2016/aug/23/virtual-reality-cool-empty-community-oculus-rift (Accessed: 21 December 2018).

Evans, L. (2019) *The Re-Emergence of Virtual Reality*. Oxon: Routledge.

Fairclough, N. (1993) 'Critical Discourse Analysis and the Marketization of Public Discourse: The Universities', *Discourse & Society*, 4(2), pp. 133–168.

Fjæstad, B. (2007) 'Why Journalists Report Science as They Do', in Bauer, M.W. and Bucci, M. (eds.) *Journalism, Science and Society: Science Communication between News and Public Relations*. Oxon: Routledge, pp. 123–131.

Fredlin, E.S. (2001) 'Framing Breaking and Creativity: A Frame Database for Hypermedia News', in Reese, S.D., Gandy, O.H. and Grant, A.E. (eds.) *Framing Public Life*. London: Lawrence Erlbaum Associates, pp. 271–295.

Fuglsang, R.S. (2001) 'Framing the Motorcycle Outlaw', in Reese, S.D., Gandy, O.H. and Grant, A.E. (eds.) *Framing Public Life*. London: Lawrence Erlbaum Associates, pp. 185–205.

Gibbs, S. (2014) 'Sony's Project Morpheus brings virtual reality to mainstream console gaming', *The Guardian*, 12 May. Available at: www.theguardian.com/tec hnology/2014/may/12/sonys-project-morpheus-virtual-reality-console-gaming (Accessed: 19 December 2018).

Go, E., Jung, E.H. and Wu, M. (2014) 'The Effects of Source Cues on Online News Perception', *Computers in Human Behavior*, 38, pp. 358–367. doi:10.1016/j.chb.2014.05.044.

Google (2017) *Glass*. Available at: https://web.archive.org/web/20170718142617/http:/www.x.company/glass/ (Accessed: 3 September 2019).

Göpfert, W. (2007) 'The Strength of PR and the Weakness of Science Journalism', in Bauer, M.W. and Bucci, M. (eds.) *Journalism, Science and Society: Science Communication between News and Public Relations*. Oxon: Routledge, pp. 215–226.

Griffiths, S. (2014a) 'From scaling heights to going shopping, a virtual reality room is now helping people with autism overcome crippling phobias', *MailOnline*, 2 July. Available at: www.dailymail.co.uk/health/article-2678267/From-scaling-heig hts-going-shopping-virtual-reality-room-helping-people-autism-overcome-crippl ing-phobias.html (Accessed: 14 February 2019).

Griffiths, S. (2014b) 'Robocop-a-feel! Droid lets you feel virtual BREASTS – and could revolutionise cancer detection', *MailOnline*, 4 June. Available at: www.dailymail.co.uk/sciencetech/article-2648536/Robocop-feel-Droid-lets-feel-virtual-BREASTS-revolutionise-cancer-detection.html (Accessed: 14 February 2019).

Griffiths, S. (2015) 'The view-master is back! Google launches £20 virtual reality device for kids that works with a smartphone', *MailOnline*, 13 February. Available at: www.dailymail.co.uk/sciencetech/article-2952468/The-View-Master-Google-launches-20-virtual-reality-device-kids-works-smartphone.html (Accessed: 14 February 2019).

Hampton, M. (2010) 'The Fourth Estate Ideal in Journalism History', in Allan, S. (ed.) *The Routledge Companion to News and Journalism*. Rev. edn. Oxon: Routledge, pp. 3–12.

Hansen, E. (2018) 'The Fourth Estate: The Construction and Place of Silence in the Public Sphere', *Philosophy and Social Criticism*, 44(10), pp. 1071–1089. doi:10.1177/0191453718797991.

Hern, A. (2016) 'Coming soon to a screen strapped to your face: virtual reality is back', *The Guardian*, 19 March. Available at: www.theguardian.com/technology/2016/mar/19/coming-soon-to-a-screen-strapped-to-your-face-virtual-reality-is-back (Accessed: 13 December 2018).

Jenner, E. (2012) 'News Photographs and Environmental Agenda Setting', *Policy Studies Journal*, 40(2), pp. 274–301. doi:10.1111/j.1541-0072.2012.00453.x.

Johnston, C (2015) 'Oculus focuses on British VR startup', *The Guardian*, 27 May. Available at: www.theguardian.com/technology/2015/may/27/facebook-oculus-british-virtual-reality-startup-surreal-vision (Accessed: 19 December 2018).

Kee, C.P., Hassan, M.A. and Ahmad, F. (2012) 'Framing the Contemporary Education Issue: Analysis of News Stories from Selected Malaysian Daily Newspapers', *Social Sciences & Humanities*, 20(2), 455–474.

Kickstarter (2012) *Oculus Rift: Step into the Game*. Available at: https://web.archive.org/web/20120801212942/http:/www.kickstarter.com/projects/1523379957/oculus-rift-step-into-the-game (Accessed: 6 September 2019).

Kristensen, N.N., Hellman, H. and Riegert, K. (2019) 'Cultural Mediators Seduced by *Mad Men*: How Cultural Journalists Legitimized a Quality TV Series in the Nordic Region', *Television & New Media*, 20(3), pp. 257–274. doi:10.1177/1527476417743574.

Krumsvik, A.H., Milan, S., Bhroin, N.N. and Storsul, T. (2019) 'Making (Sense of) Media Innovations', in Deuze, M. and Prenger, M. (eds.) *Making Media: Production, Practices, and Professions*. Amsterdam: Amsterdam University Press, pp. 193–205.

Lewis, J., Williams, A. and Franklin, B. (2008) 'A Compromised Fourth Estate?', *Journalism Studies*, 9(1), pp. 1–20. doi:10.1080/14616700701767974.

Liberatore, S. (2016) 'The HoloLens is here (if you have $3,000 to spare): Microsoft begins taking preorders from developers for augmented reality headset', *MailOnline*, 29 February. Available at: www.dailymail.co.uk/sciencetech/article-3469683/The-HoloLens-3-000-spare-Microsoft-begins-taking-preorders-developers-augmented-reality-headset.html (Accessed: 7 December 2018).

Liberatore, S. (2017a) 'Will the iPhone 8 use a radical augmented reality technology? Claims Apple has 1,000 engineers working on system', *MailOnline*, 28 February. Available at: www.dailymail.co.uk/sciencetech/article-4268778/Is-augmented-reality-Apple-s-product-innovation.html (Accessed: 11 February 2019).

Liberatore, S. (2017b) 'Samsung's strangest ad yet: Gear VR video shows an OSTRICH donning the firm's headset and learning to fly', *MailOnline*, 31 March. Available at: www.dailymail.co.uk/sciencetech/article-4366376/Samsung-uses-feathered-brand-ambassador-promote-Gear-VR.html (Accessed: 8 February 2019).

Linström, M. and Marais, W. (2012) 'Qualitative News Frame Analysis: A Methodology', *Communitas*, 17, pp. 21–28.

Lombard, M. and Ditton, T. (1997) 'At the Heart of It All: The Concept of Presence', *Journal of Computer-Mediated Communication*, 3(2). doi:10.1111/j.1083-6101.1997.tb00072.x.

Luokkanen, M., Huttunen, S. and Hildén, M. (2014) 'Geoengineering, News Media and Metaphors: Framing the Controversial', *Public Understanding of Science*, 23(8), pp. 966–981. doi:10.1177/0963662513475966.

Magic Leap (2016) 'Graeme Devine, our Chief Game Wizard, defines our playground' [Facebook] 18 April. Available at: www.facebook.com/magicleap/ (Accessed: 16 September 2019).

Magic Leap (2017) *Magic Leap*. Available at: https://web.archive.org/web/2017123 1203634/https:/www.magicleap.com/ (Accessed: 16 September 2019).

Matthes, J. and Kohring, M. (2008) 'The Content Analysis of Media Frames: Toward Improving Reliability and Validity', *Journal of Communication*, 58, pp. 258–279. doi:10.1111/j.1460-2466.2008.00384.x.

McKernan, B. (2013) 'The Morality of Play: Video Game Coverage in *The New York Times* From 1980 to 2010', *Games and Culture*, 8(5), pp. 307–329. doi:10.1177/1555412013493133.

Microsoft (2015a) *Microsoft HoloLens*. Available at: https://web.archive.org/web/20150124143645/http:/www.microsoft.com/microsoft-hololens/en-us (Accessed: 3 September 2019).

Microsoft (2015b) *Microsoft HoloLens – Transform Your World with Holograms*. 21 January. Available at: www.youtube.com/watch?v=aThCr0PsyuA (Accessed: 2 October 2019).

Microsoft (2016a) *Why HoloLens*. Available at: https://web.archive.org/web/20160301151538/http:/www.microsoft.com/microsoft-hololens/en-us/why-hololens (Accessed: 3 September 2019).

Microsoft (2016b) *Commercial*. Available at: https://web.archive.org/web/20160315185619/http:/www.microsoft.com/microsoft-hololens/en-us/commercial (Accessed: 3 September 2019).

Microsoft (2017) *Commercial Overview*. Available at: https://web.archive.org/web/20170412022555/http:/www.microsoft.com/en-us/hololens/commercial-overview#EditorialPivotMainBlockFocus (Accessed: 3 September 2019).

Müller, M.G., Kappas, A. and Olk, B. (2012) 'Perceiving Press Photography: A New Integrative Model, Combining Iconology with Psychophysiological and Eye-Tracking Methods', *Visual Communication*, 11(3), pp. 307–328. doi:10.1177/1470357212446410.

Nordfors, D. (2009) 'Innovation Journalism, Attention Work and the Innovation Economy', *Innovation Journalism*, 6(1), pp. 1–46.

Norris, P. and Odugbemi, S. (2010) 'Evaluating Media Performance', in Norris, P. (ed.) *Public Sentinel: News Media & Governance*. Washington, DC: The World Bank, pp. 3–29.

Oculus (2015a) *Oculus Rift Reveal – Step into the Rift*. 11 June. Available at: https://youtu.be/etv_IxVh7cc (Accessed: 6 September 2019).

Oculus (2015b) 'What's the first thing you show to introduce family and friends to virtual reality using the Gear VR?' [Facebook] 7 December. Available at: www.facebook.com/Oculusvr (Accessed: 16 September 2019).

Oculus (2016) 'The newest Samsung Gear VR, powered by Oculus, is available today alongside the Galaxy Note 7!' [Facebook] 19 August. Available at: www.facebook.com/Oculusvr (Accessed: 16 September 2019).

Oculus (2017a) *Oculus Rift | Step into Rift — now only $399*. 11 October. Available at: https://youtu.be/5q6BcQq_yhw (Accessed: 6 September 2019).

Oculus (2017b) 'There's more to VR than just epic games' [Facebook] 20 August. Available at: www.facebook.com/OculusGB (Accessed: 3 April 2020).

Oculus (2017c) 'Transport yourself to Japan's spectacular Motenashi Dome of Kanazawa Station in 360° on Rift and Gear VR' [Facebook] 12 December. Available at: www.facebook.com/Oculusvr (Accessed: 21 November 2019).

Oculus VR (2012) *New Virtual Reality Gaming Headset from Oculus Gets Kickstarted*. 1 August [Press Release]. Available at: https://web.archive.org/web/20120811020422/http://oculusvr.com:80/press_release/ (Accessed: 21 December 2018).

Oculus VR (2015a) *Rift*. Available at: http://web.archive.org/web/20150611180844/https://www.oculus.com/en-us/rift/ (Accessed: 2 September 2019).

Oculus VR (2015b) *Gear VR*. Available at: https://web.archive.org/web/20151127171834/https://www.oculus.com/en-us/gear-vr/ (Accessed: 2 September 2019).

Oculus VR (2015c) *Oculus*. Available at: http://web.archive.org/web/20150709170214/https://www.oculus.com/en-us/ (Accessed: 30 August 2019).

Oculus VR (2016) *Rift*. Available at: http://web.archive.org/web/20161008154031/https://www3.oculus.com/en-us/rift/ (Accessed: 3 September 2019).

Plummer, L. and O'Hare, R. (2016) 'PlayStation VR is now on sale: Sony's virtual reality gaming headset will take on the Oculus Rift and HTC Vive', *MailOnline*, 12 October. Available at: www.dailymail.co.uk/sciencetech/article-3834393/Sony-tapping-virtual-reality-PlayStation-headset.html (Accessed: 11 February 2019).

Prigg, M. (2014a) Facebook buys virtual reality headset firm Oculus for $2bn as Mark Zuckerberg promises to 'change the way we communicate', *MailOnline*, 25 March. Available at: www.dailymail.co.uk/sciencetech/article-2589367/Get-ready-social-platform-Facebook-buys-virtual-reality-firm-Oculus-2bn.html (Accessed: 14 February 2019).

Prigg, M. (2014b) 'Google glass for war: the US military funded smart helmet that can beam information to soldiers on the battlefield', *MailOnline*, 27 May. Available at: www.dailymail.co.uk/sciencetech/article-2640869/Google-glass-war-US-military-reveals-augmented-reality-soldiers.html (Accessed: 13 February 2019).

Richardson, J.E. (2007) *Analysing Newspapers*. Basingstoke: Palgrave.

Roderick, I. (2016) *Critical Discourse Studies and Technology*. London: Bloomsbury.

Rogers, E.M. (2003) *Diffusion of Innovations*. 5th edn. New York: Free Press.

Ryan, M.L. (2015) *Narrative as Virtual Reality 2*. Baltimore: Johns Hopkins University Press.

Sheafer, T., Shenhav, S.R. and Amsalem, E. (2018) 'International Frame Building in Mediated Public Diplomacy', in D'Angelo, P. (ed.) *Doing News Framing Analysis II*. Oxon: Routledge, pp. 249–273.

Shen, F. (2004) 'Chronic Accessibility and Individual Cognitions: Examining the Effects of Message Frames in Political Advertisements', *Journal of Communication*, 54(1), pp. 123–137. doi:10.1111/j.1460-2466.2004.tb02617.x.

Shoemaker, P.J. and Reese, S.D. (2014) *Mediating the Message in the 21st Century*. Oxon: Routledge.

Shubber, K. (2014) 'Are you ready for the virtual reality revolution?', *The Guardian*, 2 August. Available at: www.theguardian.com/technology/2014/aug/02/are-you-ready-for-virtual-reality-revolution-google-cardboard (Accessed: 13 December 2018).

Spettigue, S. (2015) "Intuitive controls and gorgeous visuals': Dailymail.com goes 'heads on' with Sony's PlayStation VR as experts predict it could outsell Facebook's Oculus Rift', *MailOnline*, 28 October. Available at: www.dailymail.co.uk/sciencetech/article-3292877/Dailymail-com-goes-heads-Sony-s-PlayStation-VR-experts-predict-outsell-Facebook-s-Oculus-Rift.html (Accessed: 13 February 2019).

Steptoe, W., Julier, S. and Steed, A. (2014) 'Presence and Discernability in Conventional and Non-Photorealistic Immersive Augmented Reality', *Proceedings of IEEE International Symposium on Mixed and Augmented Reality*, Munich, Germany, 10–12 September, pp. 213–218.

Stuart, K. (2015) 'Minecraft on Hololens: the future of gaming is right in front of your eyes', *The Guardian*, 24 June. Available at: www.theguardian.com/technology/2015/jun/24/minecraft-hololens-mixed-augmented-reality-microsoft (Accessed: 19 December 2018).

Tankard, J.W. (2001) 'The Empirical Approach to the Study of Media Framing', in Reese, S.D., Gandy, O.H. and Grant, A.E. (eds.) *Framing Public Life*. Oxon: Routledge, pp. 95–105.

Taylor, J. (1997) 'The Emerging Geographies of Virtual Worlds', *The Geographical Review*, 87(2), pp. 172–192.

Van Gorp, B. (2007) 'The Constructionist Approach to Framing: Bringing Culture Back In', *Journal of Communication*, 57, pp. 60–78. doi:10.1111/j.1460-2466.2006.00329.x.

Van Gorp, B. (2010) 'Strategies to Take Subjectivity Out of Framing Analysis', in D'Angelo, P. and Kuypers, J.A. (eds.) *Doing News Framing Analysis*. Oxon: Routledge, pp. 84–109.

Whitton, N. and Maclure, M. (2015) 'Video Game Discourses and Implications for Game-based Education', *Discourse: Studies in the Cultural Politics of Education*, pp. 1–13. doi:10.1080/01596306.2015.1123222.

Williams, D. (2003) 'The Video Game Lightning Rod', *Information, Communication & Society*, 6(4), pp. 523–550. doi:10.1080/1369118032000163240.

Woollaston, V. (2015) 'Virtual reality for the masses: £30 immerse headset lets you journey through VR worlds and watch 3D movies on your phone', *MailOnline*, 16 April. Available at: www.dailymail.co.uk/sciencetech/article-3041609/Virtual-reality-masses-30-Immerse-headset-lets-journey-VR-worlds-watch-3D-movies-phone.html (Accessed: 13 February 2019).

Zolfagharifard, E. (2014) 'The contact lenses that could do away with TV screens: system that projects images onto the eyeball to be unveiled next week', *MailOnline*, 3 January. Available at: www.dailymail.co.uk/sciencetech/article-2533463/The-contact-lenses-away-TV-screens-System-projects-images-eyeball-unveiled-week.html (Accessed: 12 December 2018).

5 Framing XR as an Innovation
Newness and Quality

Krumsvik et al. state that "innovation implies introducing (and implementing) something *new* into the socioeconomic system" (2019: 194, emphasis added). In other words, this newness is one of the major features of innovations such as XR. The second frame category, Newness, allows researchers to examine how this concept has been presented in media coverage of an emerging technology. It focuses on identifying frames that highlight what makes a technology new or different. Researchers can ask a number of questions when analysing the sample to identify such frames:

- What is said to be significantly new/different about this technology compared to others?
- What new experiences/opportunities is the technology said to afford?
- What impact or changes is the technology said to bring?
- How advanced (or not) is the technology said to be?

The study presented in this book found that three of the frames applied to XR link to the newness of the technology: Different and Unique; Revolutionary and Transformative; and Advanced and High-Quality. Each of these frames could be identified in both the news and marketing texts. Following the same format as the previous chapter, this chapter examines the framing devices used to construct these frames and makes comparisons between the news and marketing.

Different and Unique

The characteristics of an innovation that make it different from existing technologies give it an element of newness, since it does something that previous innovations did not. A Different and Unique frame was applied to XR during its inception, both in the news coverage and marketing. Broadly, this involved portraying XR (either the technology or applications) as different or unique to other technologies/forms of media.

DOI: 10.4324/9781003375814-5

Firstly, as certain words can act as framing devices (Entman, 1993; Linström and Marais, 2012), the existence of this frame in the news articles can be indicated by the use of words in the "different and unique" category. Such terms appeared 112 times in 9 percent of news articles overall. However, in a different way to the Immersive and Transcendent frames discussed in the previous chapter, this frame was not similarly prevalent in all news outlets. Whereas both the *Sun* and the *Guardian* used these words in a comparable portion of articles (15 and 14 percent respectively), the *MailOnline* included "different and unique" words in just 6 percent of its articles. While these figures show that each news outlet used words that highlight the Different and Unique frame, it appears that the strength of the frame has been affected by the media organisation writing about XR, providing insight into the impact of this factor of the hierarchy of influences model (Shoemaker and Reese, 2014).

On the other hand, in a similar way to the Immersive and Transcendent frames, the type of XR the news articles focused on impacted how often this frame was used. Words in the "different and unique" category appeared in 13 percent of articles focusing on VR compared to 7 percent of those about AR/MR. Therefore, the frame-building process appears to have been influenced by the technological characteristics of the devices being written about in the news. This finding also helps to explain the low percentage of articles using words in the "different and unique" category in 2012 and 2013, since these years focused predominantly on AR/MR. After 2013, the use of terms in the "different and unique" category ranged from a high of 16 percent in 2014 to a low of 11 percent in 2017. Though this shows a slight fluctuation, these figures suggest that the Different and Unique frame appeared relatively consistently between 2014 and 2017. As mentioned in the previous chapter, the frame repetition hypothesis argues that "the frames that have greater influence on public opinion are those that are repeated more frequently" (Sheafer, Shenhav and Amsalem, 2018: 264). As a result, the repetition of the Different and Unique frame across multiple years could have a significant impact on how readers view XR.

In more detail, examining the individual words within this category shows that the stem *unique** was used the most (7 percent), closely followed by *different* (4 percent). Alternatively, *weird** was only used in 2 percent of articles. For *unique** to be the most common term within this category demonstrates that the news articles have not only portrayed XR as different from other forms of media, but that they have moved beyond this to suggest the technology is unlike anything that has come before it. In a study of 177 businesses creating new products, Cooper found that the "single most important dimension leading to new product success is Product Uniqueness and Superiority" (1979: 100). More recently, Flight et al. linked "uniqueness of features" (2011: 110) to the perceived characteristic of relative advantage as one factor supporting innovation diffusion. Thus, for *unique** to be the most used term in this category suggests the news articles could promote the diffusion of XR. Furthermore, it is significant that *weird** was rarely used within the news articles since this term typically has more negative connotations than *different* and *unique**. It appears that word choice, as a framing device, has contributed to framing XR as Different and Unique, specifically in a positive light.

Additionally, across the sample, news articles framed XR as Different and Unique by describing several specific devices as the first of their kind. The following examples demonstrate this:

the world's first really viable virtual reality headset

(Poole, 2014)

World's first true augmented reality ski goggles

(Sturgis, 2015)

Fove is the first virtual reality headset to use eye-tracking technology

(Williams, 2015)

Samsung has made history of a sort by launching the first major consumer-oriented virtual-reality headset

(Associated Press, 2015)

Whether broadly in relation to VR (Poole, 2014), for a specific scenario (Sturgis, 2015) or regarding the technology the devices use (Associated Press, 2015; Williams, 2015), each product is portrayed as being unique because it is said to be the first to offer something. This acts as a rhetorical framing device in the construction of the Different and Unique frame. Moreover, these examples each highlight the use of technical framing devices to increase the salience of the frame. To expand, the second excerpt (Sturgis, 2015) appeared as the headline of the article, the first (Poole, 2014) and fourth (Associated Press, 2015) both appeared in the lead paragraphs of their respective articles and the third excerpt (Williams, 2015) was the top bullet point in the article summary that the *MailOnline* includes at the beginning of its news items. News articles are typically structured using the inverted pyramid design in which elements are placed "in decreasing order of importance or newsworthiness" (Bednarek and Caple, 2012: 100). Pan and Kosicki claim that the headline of an article is "the most powerful framing device of the syntactical structure", while the "lead is the next most important device to use" (1993: 59). In other words, any points that appear in the headline and lead paragraph of articles are given particular emphasis. Therefore, the Different and Unique frame has featured in prominent positions within these examples, demonstrating its strength.

Considering the prominence of the idea that these products are the first of their kind, it is significant that the same framing device was used in XR marketing. Promotional materials for both Oculus Rift and HoloLens highlighted the Different and Unique frame this way. Firstly, the Oculus Rift Kickstarter page claimed the device was "the first truly immersive virtual reality headset for video games" (Kickstarter, 2012). The press release for the campaign also stated that users will be able "to experience VR gaming for the first time" (Oculus VR, 2012). Secondly, the HoloLens website, as well as a promotional video, defined the device as "the world's first fully untethered, self-contained holographic computer" (Microsoft, 2015a, 2016d). This shows that the Different and Unique frame appeared in both samples, highlighting a further similarity between the two discourses. Not only that, but the same rhetorical framing device (the idea of the products being the first of their kind) was used in both the news and marketing. Again, it appears that these two texts work together to reinforce this frame to the public.

In a similar way, both discourses also constructed the Different and Unique frame by representing the XR experience as unlike anything else. For instance, in the news, a *Guardian* article claimed that "VR offers the potential to put the viewer into the experience of actually being there like nothing else" (Gibbs, 2016). Here, the idea of immersion ("putting the viewer into the experience") is used to argue that the experience is unique. In XR marketing, a similar phrase was used as a framing device to imply the same sentiment: "like never before". This phrase appeared in marketing of Gear VR, HoloLens and Oculus Rift, as demonstrated below:

Gear VR is a virtual reality headset that lets you experience games, movies and more like never before

(Samsung US, 2015)

Immerse yourself in entertainment like never before

(Samsung, 2016)

Connect, create, and explore like never before

(Microsoft, 2015a)

See detail like never before when you build in 3D

(Microsoft HoloLens, 2016)

explore new worlds like never before

(Kickstarter, 2012)

dive into the Blade Runner universe like never before

(Oculus, 2017a)

With the phrase appearing across the marketing of multiple devices, this is further evidence to indicate that the Different and Unique frame is salient in marketing discourse. The persuasive power of frames is enhanced when they appear in more than one type of media (Van Gorp, 2007). Therefore, the news has reinforced the same frame from the marketing that is intended to sell the products, thus potentially supporting its adoption.

However, unlike the frames previously discussed, one article in the *Guardian* attempted to counter the Different and Unique frame. The article, headlined "Facebook's virtual reality [Oculus Rift] just attempts what artists have been doing forever" (Judge, 2016), cites Oculus' Mark Zuckerberg but criticises his argument. The sub-heading of the report stated: "Mark Zuckerberg says VR will capture human experiences like never before – but is it really superior to what writers and artists achieved centuries ago?" (Judge, 2016). The article went on to argue that virtual worlds have been created by writers and artists for "centuries"; thus, Oculus Rift is nothing new or unique. Nevertheless, this is the only instance that could be identified of the Different and Unique frame being contested. To reiterate, Van Gorp states that the more often frames "confirmed by further information, or by congruent framing devices, the more difficult it becomes to refute or change them by counterframing" (2007: 69). Since the Different and Unique frame appeared in both the news and marketing discourses and multiple framing devices were used to do this, it is unlikely that the one attempt at counterframing will have much effect on readers.

Deviance is a common focus of moral panics (Cohen, 2002), in which actors (including news media) highlight issues over something that is different to the norm. With this in mind, it might be expected for the aspects of XR that are different from other technologies to be portrayed in such a way to create a moral panic. However, this has not been the case for XR. Although the characteristics of the technology that set it apart from others are highlighted, this is done in a positive way through the Different and Unique frame to present XR as new and appealing to readers. News about XR therefore differs from other technologies that have been the subject of a moral panic, including radio, TV (Markey and Ferguson, 2017), mobile phones (Goggin, 2006) and videogames (Rogers, 2013).

Alternatively, the appearance of the Different and Unique frame in the news coverage shows some similarity with Therrien and Lefebvre's (2017) study of videogame marketing. To expand, Therrien and Lefebvre found that videogame marketing often includes "the classic 'old vs. new' antithesis", which is characterised by "expressions such as 'the first of'" a particular technological novelty (2017: 62). This coincides with the finding that XR was framed as Different and Unique by claiming it is the first of its kind. As mentioned above, the uniqueness of a product makes it more likely to be adopted (Cooper, 1979; Flight et al., 2011). Therefore, framing XR as Different and Unique could arguably promote its diffusion by emphasising these supposedly unique features. Furthermore, since one of the aims of marketing is to differentiate a product from others to highlight its value to consumers (Kotler et al., 2016), it is not surprising that the promotional materials for these devices have highlighted their uniqueness. However, as this is one of the main aims of product promotion, the fact that the news articles also represent XR as unique means that they are potentially aiding the marketing of XR products. This is further evidence of the marketisation (Fairclough, 1993) of XR news and the blurring boundary between news and promotional content.

Revolutionary and Transformative

Related to Different and Unique, another frame that appeared in both the news and marketing of XR was Revolutionary and Transformative. A technological revolution can be defined as "a dramatic change brought about relatively quickly by the introduction of some new technology" (Bostrom, 2007). Certainly, in news and marketing of XR, the Revolutionary and Transformative frame involved presenting XR as a technology that could radically change certain areas.

As before, the use of terms in the "revolutionary and transformative" category help to illustrate how often this frame was used. More common than terms in the "different and unique" category, words in the "revolutionary and transformative" group appeared in 16 percent of articles overall. The *Guardian* was most likely to use such words, with 18 percent of articles from this publication including terms in the "revolutionary and transformative" category. The *MailOnline* used these terms slightly less (15 percent). However, the *Sun* only used words from the "revolutionary and transformative" category in 8 percent of its articles. This indicates that the frame was fairly prominent in the *Guardian* and the *MailOnline*, though it was not used very often in the *Sun*. The media organisation factor of the hierarchy of influences model (Shoemaker and Reese, 2014) appears to have impacted the strength of this frame, though not to the extent that it was absent from some news outlets. This shows similarities with the Different and Unique frame.

Furthermore, there was only a slight difference between the use of words highlighting this frame in articles about VR and those focusing on AR/MR. Of those articles focusing on VR, 15 percent included words in the "revolutionary and transformative" category. Similarly, 19 percent of articles concentrating on AR/MR used such terms. Therefore, this frame appears to have been applied to AR/MR slightly more often than VR. However, the difference is fairly small, showing that the technological characteristics of the devices do not appear to have impacted the prevalence of this frame to a great extent.

Alternatively, the use of words construing this frame varied to a slightly larger degree in the different years of the sample period. In 2012, 21 percent of articles included terms from the "revolutionary and transformative" category. This shows that when XR products were first announced in 2012, it was common for the news outlets to frame them as Revolutionary and Transformative. However, the frame was not as prominent in 2013, as uses of these words dropped to 12 percent. On the other hand, Mark Zuckerberg's involvement with XR through the purchase of Oculus appears to have renewed the use of the Revolutionary and Transformative frame, since 22 percent of articles used words in this category in 2014. Still, after 2014, this figure gradually decreased until the end of the sample period (2017) when it ended at 11 percent. This shows that the use of this frame has not been as stable as others. Nevertheless, words in the "revolutionary and transformative" category were still present every year, which, according to the frame repetition hypothesis (Sheafer, Shenhav and Amsalem, 2018), could increase its strength.

Examining the use of specific terms within the "revolutionary and transformative" category provides further insight into the construction of this frame. Out of all words in this category, the stem *revolution** was used the most, in 9 percent of articles. The stem *transform** appeared the second most, being present in 6 percent of articles. On the other hand, the word *disruptive* was rarely used, appearing just ten times overall in 1 percent of news items. As *disruptive* has more negative connotations than *revolution** or *transform**, this highlights that the supposedly revolutionary aspects of XR have not been presented in a negative light. Moreover, these results share similarities with the preference for *different* and *unique** over *weird** mentioned in the previous section. Van Dijk highlights that the choice of specific words can indicate "the attitudes and hence ideologies of the speaker" (1988: 81). In relation to the frames discussed in this chapter thus far, these specific word choices indicate that the news media present favourable attitudes towards XR.

As well as word choice, several additional framing devices were used to present XR as Revolutionary and Transformative. Firstly, exemplars were used as framing devices in the news to relate XR's supposedly revolutionary and transformative capabilities to a wide range of areas. For instance, a *MailOnline* article mentioned that Magic Leap "could revolutionise how people shop, watch TV and even how doctors operate" (Griffiths and Prigg, 2015). Likewise, an article from the *Guardian* noted that: "Games, exploration, psychiatry and many other fields could all be revolutionised" by VR (Arthur, 2015). In these sentences, the revolutionary potential for the technology is presented as broad due to the range of areas mentioned. Desrosiers argues that frames resonate with the public "when they reflect what publics live, what they believe, and what they believe matters" (2012: 5). Therefore, relating the Revolutionary and Transformative frame to a wide range of areas increases the portion of readers that this will be relevant to, thus making it resonate with a wider audience and increasing the salience of the frame overall.

In a similar way, the phrase "future of" was used as a rhetorical framing device to construct the Revolutionary and Transformative frame. For example, a *MailOnline* article stated that a PlayStation VR demo "clearly demonstrated the potential of the technology for the future of home entertainment" (Spettigue, 2015). Claiming that the technology could be the "future of" something implies that it is not only different to what came before it (linking to the Different and Unique frame above), but that this difference will be an improvement to the extent that it will replace its predecessor. In other words, it will change, transform or revolutionise that area. Moreover, in other articles, this technique was not restricted to XR's impact in entertainment. Firstly, a *MailOnline* article highlighted that Google Glass "has been touted as the future of computing" (Prigg, 2013). While this sentence only notes one area of impact, computing in itself is very broad. Therefore, describing Google Glass as the future of computing suggests its revolutionary potential could be very meaningful. In a different way, an article in the *Guardian* focusing on PlayStation VR stated that the device provides "a convincing look at the future of gaming, virtual tourism and a whole new set of experiences" (Gibbs, 2014). Here, as in the examples discussed previously, this presents XR as affecting an assortment of areas which could make this frame resonate with a larger audience (Desrosiers, 2012).

Considering the relationship between the news and marketing, it was found that the same phrase ("future of") was used as a framing device in XR promotional content. Specifically, Oculus Rift marketing claimed that the device is the future of gaming (Facebook, 2015; Oculus, 2017b). More broadly, promotional material for both Magic Leap and HoloLens claimed they were creating the future of computing (Magic Leap, 2017; Microsoft, 2016c, 2016d), with HoloLens marketing sometimes more precisely referring to it being the future of *holographic* computing (Microsoft, 2015a, 2015d, 2016a). Furthermore, HoloLens promotional texts also argued that the device will be the future of product design (Microsoft, 2016a, 2016b, 2016e), architecture, learning and education, construction and home improvement (Microsoft HoloLens, 2016). In Magic Leap marketing, this idea was extended, with a press release stating that the device is "not just the future for entertainment; it has the potential to be the future of everything" (Magic Leap, 2016a). These examples demonstrate that both discourses have used the "future of" phrase to highlight the revolutionary and transformative potential of XR, highlighting further overlap between them. Again, a frame that is used to promote XR has also appeared in the news. This reinforces the frame and the likelihood of it being accepted by readers (Van Gorp, 2007), raising questions as to the separation of news and promotional content in XR coverage.

As well as mentioning a range of areas that will be revolutionised by XR, a more specific exemplar was used as a framing device when focusing on the Magic Leap headset. For example, one *MailOnline* article headline read as follows: "Magic Leap set to revolutionise every aspect of daily life" (Griffiths and Prigg, 2015). Here, instead of referencing a specific area that will be revolutionised by XR, the Magic Leap device itself is described as revolutionary. As this frame appears in the headline of the article, it is particularly salient (Pan and Kosicki, 1993). The frame is also enhanced by the claim that it can revolutionise *every* aspect of life. In a similar way, the *Guardian* stated that: "some say [Magic Leap] may be the most revolutionary tech gadget in years" (Yadron, 2016). The modifier "most" suggests its revolutionary capabilities are very strong. Although the lack of a specific source for this point could negatively impact its perceived credibility for readers (Duffy and Freeman, 2011), it nevertheless holds more weight than if the journalist had presented it as their own opinion. These examples demonstrate that the Revolutionary and Transformative frame is particularly strong in relation to Magic Leap.

This is a significant finding when considered alongside the marketing of Magic Leap. Even the name and logo of the company frame the technology as Revolutionary and Transformative. A press release about Magic Leap's first round of funding clearly highlighted this. Stryker, Magic Leap's parent company, stated that the technology is "truly game changing. It is like a rocket ship for the mind" (Magic Leap, 2014). Firstly, calling Magic Leap "game changing" suggests that it will transform the industry. This is emphasised with the rocket ship metaphor, echoing Magic Leap's logo which is itself a rocket. Since the rocket ship was revolutionary for space exploration, this implies Magic Leap will be revolutionary for "the mind", hinting at great potential for seemingly limitless areas. Furthermore, in a Facebook post (Magic Leap, 2016b), Magic Leap states that their company name was inspired by the moon landing ("one giant *leap…*"). In addition to the company's name and logo, the idea that Magic Leap is transformative was highlighted on their website. When inviting developers to consider how Magic Leap could be used, the website stated: "Imagine how this would completely transform how people interact with both the digital and real-worlds. Imagine you being one of the first to help transform the world forever" (Magic Leap, 2015). In this quote, the use of "completely" to describe Magic Leap's transformative capabilities implies that it will result in highly substantial change. The statement also stresses that this will be a lasting change because it will "transform the world forever", which makes its revolutionary and transformative capabilities appear even stronger. Therefore, based on Magic Leap's overall branding, as well as statements such as this, it appears the company is a strong advocate of the Revolutionary and Transformative frame. With this in mind, the *MailOnline* and the *Guardian*'s references to Magic Leap as the *most* revolutionary, with the potential to transform every aspect of life, shares a very similar sentiment to Magic Leap's overall marketing. Combined with the appearance of other shared frames, this finding indicates that, within the social institutions factor of the hierarchy of influences (Shoemaker and Reese, 2014), the marketing has impacted the frame-building of XR in the news.

Further evidence of this can be identified in the news articles' use of XR company owners as sources to portray XR as Revolutionary and Transformative. An example is a quote from Mark Zuckerberg. The statement itself originated from a press release announcing Facebook's plans to acquire Oculus. In the press release, Zuckerberg claimed: "Oculus has the chance to create the most social platform ever, and change the way we work, play and communicate" (Facebook, 2014). Of particular interest in relation to the Revolutionary and Transformative frame is the idea that the device can "change the way we work, play and communicate". Zuckerberg implies that Oculus Rift will change nearly every aspect of life (work, play and communication), making its transformative capabilities seem extremely far-reaching. Press releases can be considered as information subsidies, or "an attempt to produce influence over the actions of others by controlling their access to and use of information relevant to those actions" (Gandy, 1982: 61). They are used by organisations with the aim of conveying their desired message within the news (Hecht et al., 2017: 740). In this case, Zuckerberg's quote appeared in three different articles in the *Guardian* and 18 *MailOnline* articles, indicating that information subsidies have been effective in getting a frame advocate's point into the news. Information subsidies relate to the social institutions factor of the hierarchy of influences model (Shoemaker and Reese, 2014), showing the impact of this factor on the frame-building process. Moreover, the routine practice (the fourth factor) for the *MailOnline* to copy and paste information from one article to another has strengthened this frame.

Furthermore, the fact that any quotations have appeared in multiple articles within the same outlet suggests that journalists are under pressure to create news content quickly, particularly in the *MailOnline* where this happens the most. Bauer et al. (2013) found that almost two thirds of science journalists "recognise that working pressures are harming the quality of science stories" (2013: 29; see also Schäfer, 2017). Similarly, Williams and Clifford's (2009) study found that 46 percent of science and technology journalists said they have less time to fact-check stories than they did previously and 22 percent noted they no longer had time to properly fact-check stories, thus damaging news quality. Because of this, journalists have become increasingly dependent on easily accessible and pre-packaged sources, such as press releases. This result in XR news coverage suggests that the commercial pressures within the social system have resulted in this routine practice that speeds up the creation of content, linking these two factors of the hierarchy of influences model. Not only does this lessen the quality of the news, but when quotations are copied and pasted, it provides those sources with significant power. In this case, Zuckerberg has been given considerable power in defining XR, which has increased the salience of the Revolutionary and Transformative frame.

Comparably, another article used a quotation from Apple CEO Tim Cook to highlight this frame in relation to AR, showing similarities with the Transcendent frame. The *MailOnline* article read as follows: "When asked what technologies he sees as transformative, Cook said: 'I'm incredibly excited by AR because I can see uses for it everywhere'" (Best, 2017b). Using the same technique as the quotes above that have related XR's revolutionary and transformative impact to a wide range of areas, Cook argues that AR can be used "everywhere". This, again, makes AR's transformative potential seem strong. Apple is a very established and successful brand, thought to be at the technological high-end. Indeed, as of 2022, Apple has been at the top of *Forbes*' list of the world's most valuable brands for ten consecutive years, being worth $482,215 million USD (Faithfull, 2022). Additionally, Apple (2018) reported that the number of active users of their devices reached 1.3

billion as of January 2018 (approximately three months after this news article was published). It is clear the company has a very large user base and is financially successful. Because of this, Cook's opinion could be deemed as highly credible for readers. According to van Dijk, discourse is more persuasive when journalists select "reliable, official, well-known, and especially credible persons and institutions" as sources (1988: 94). Therefore, quoting Cook giving this opinion increases the salience of the Revolutionary and Transformative frame for AR in particular. Moreover, within the social institutions factor (Shoemaker and Reese, 2014), these two sources (Zuckerberg and Cook) have acted as advocates of this frame.

In a different way to the framing devices already discussed, another technique was used in HoloLens marketing to present XR as Revolutionary and Transformative. Instead of simply claiming the device can revolutionise an area, HoloLens was presented as a device that can allow its users to "transform the world". For instance, the tagline of the product is "transform your world", which appeared on their website, at the end of their promotional videos and on their social media pages (e.g. Microsoft 2015a). This tagline in itself implies that the device can be used as a transformative tool. Additionally, the video advert for HoloLens titled "Transform your world with holograms" further highlighted this when the narrator stated: "when you change the way you see the world, you can change the world you see" (Microsoft, 2015b). HoloLens is represented as transformative in the sense that it can facilitate transformation through its use, rather than the device itself being inherently revolutionary or transformative. As this idea of transforming the world is central to HoloLens marketing, this further demonstrates the prominence of the frame in XR promotional materials. For this frame to appear in both the news and marketing means it is more likely to be accepted as reality (Van Gorp, 2007), thus demonstrating the significance of this frame being shared between the two discourses.

Although the Revolutionary and Transformative frame is salient within the news and marketing, it is important to note that one *MailOnline* article used a statement from a "senior research analyst" arguing against it (Plummer and O'Hare, 2016). The source argued that: "Instead of being a game-changer, VR is likely to give a boost to the gaming industry" (Plummer and O'Hare, 2016). This limits VR's impact to just one area (gaming) and suggests that its effects will not be substantial enough to be classed as revolutionary or transformative. Pan and Kosicki (1993) highlight that the designator given to a source (i.e. how a source is labelled) can impact the authoritativeness of the statement. In this case, naming the source a *senior* research analyst gives them, and the statement, a strong level of authoritativeness. However, the prominence a journalist gives to a quotation is also part of the framing process (Van Gorp, 2010). Out of the 25 paragraphs in the article, this statement appeared in the 22nd and 23rd. In other words, according to the inverted pyramid structure, it is not in a prominent position. Thus, although the designator applied to this source has given them a certain credibility, the placement of this statement has significantly lessened its emphasis. Considering that some of the examples above have highlighted the Revolutionary and Transformative frame within article headlines, it is clear the sample has favoured the Revolutionary and Transformative frame rather than any sources highlighting counterframes.

The appearance of the Revolutionary and Transformative frame in the news and marketing discourses shows continuity with Chan's finding that literature and film in the 1990s presented VR as "revolutionary and unprecedented" (2014: 124). Extending past XR, both Roderick's (2016) broad study of technology discourse and Kelly's (2009) study of magazine coverage of microcomputers uncovered representations of technology as revolutionary. In this way, XR appears to have been framed similarly to other technologies. Moreover, framing XR as Revolutionary and Transformative presents the technology as having substantial importance since it will supposedly have particularly meaningful implications. Regarding the diffusion of innovations, Rogers (2003) argues that the higher the perceived importance of an innovation, the more likely it is to be adopted. Therefore, framing XR in this way could arguably promote its diffusion. This is particularly the case since the frame appears in both news and marketing discourse, which work to reinforce each other and thus the facticity of the frame (Van Gorp, 2007). These results coincide with the findings regarding the frames that have previously been discussed.

Similarly to the Different and Unique frame, the revolutionary and transformative capabilities of XR could have been presented as deviant to create a moral panic. Again, this did not happen. Instead of suggesting the change XR can bring about is disruptive, the news articles have presented these effects positively. This further shows that the news outlets have favoured positive framings of XR which can work towards supporting XR marketing. As a result, the news works in the commercial interests of the companies selling XR products.

Advanced and High-Quality

The final frame to be discussed in this chapter is Advanced and High-Quality. This frame involved presenting XR technology as advanced and thus capable of providing a high-quality XR experience. The use of this frame contributes to the broader concept of newness by highlighting the advanced nature of the products; for something to be advanced suggests it is utilising some of the newest technology and features currently available. As was found in the previous sections, a range of framing devices were used in the news and marketing texts to construct the Advanced and High-Quality frame and this segment analyses them in detail.

To begin, quantitative data from the frequency of terms analysis shows that the Advanced and High-Quality frame was particularly prominent in the news articles. Across the entire news sample, words in the "advanced and high-quality" category appeared 500 times in 30 percent of articles. At almost one third of the news sample, this is a substantial amount. Furthermore, out of all word categories corresponding to a specific frame, those in the "advanced and high-quality" group were used in the second largest portion of articles, after those referring to the Immersive frame. Although a frame does not always have to be repeated often to have an effect on the public (Entman, 2003; Van Gorp, 2010), "repetitive framing will strengthen a news framing effect" (Lecheler and de Vreese, 2019: 88). Therefore, the large portion of articles with words that refer to XR as Advanced and High-Quality show that this is a particularly strong frame within the news.

However, there were some slight variations between the news outlets in the use of the Advanced and High-Quality frame. Words in the "advanced and high-quality" category were used in the second largest portion of articles in the *Sun* and *MailOnline*, after those from the "immersive" group. On the other hand, the category of words that was used the second most in the *Guardian* was "much-anticipated", with the "advanced and high-quality" category ranking third. Still, the difference in percentage was quite small (see Appendix 5). This shows that, regarding the hierarchy of influences model (Shoemaker and Reese, 2014), the media organisation reporting on XR has had an impact on the strength of the Advanced and High-Quality frame, though only to a small degree. When information is corroborated by more than one source, audiences are more inclined to accept it as true (Nordfors, 2009). Thus, for the Advanced and High-Quality frame to be present in all three news outlets increases its potential strength in impacting public opinion.

Furthermore, there was only a slight difference in the use of words in the "advanced and high-quality" category when referring to VR or AR/MR. Across the entire sample period, 28 percent of VR articles used words from this category, whereas 35 percent of AR/MR articles included these terms. This shows that the characteristics of these technologies has had little impact on whether or not this frame was used. Still, the fact that slightly more AR/MR articles included words from this category helps to explain the peak of the use of these terms in 2013, a year in which 70 percent of articles focused on AR/MR. Aside from the peak in 2013, the use of words in the "advanced and high quality" category remained fairly consistent across the years. This did decrease year-on-year from 2014, though not for a substantial amount. Despite minor variations, this data indicates that the Advanced and High-Quality frame was prominent in every year of the sample. Since the repetition of frames over time can increase their influence on public opinion (Dickerson, 2001; Sheafer, Shenhav and Amsalem, 2018), the consistent appearance of this frame could lead readers to believe XR to be advanced and high-quality.

In addition to the appearance of certain words, other framing devices were used to construct the Advanced and High-Quality frame. A major part of this was referencing fiction. This included stating real XR products were similar to fictional depictions of XR. Several news article headlines used this technique:

Inception helmet creates alternative reality

(Costandi, 2012)

Google glasses with built-in **Terminator-style** computer displays "could be on sale by the year's end at a cost of $250"

(Waugh, 2012)

Generals will be able to direct battles using new **Minority Report-style** technology including 3D goggles and even virtual reality contact lenses

(Burrows, 2015)

"It was a bit **like the Matrix**": FIFO father becomes world's first man to experience son's birth from 4000km away after breakthrough in virtual technology

(Carty, 2015)

Star Wars-style moving holograms are here: Microsoft shows how HoloLens can bring distant family members into your home

(Macdonald, 2016)

Could virtual reality prevent depression in ASTRONAUTS? **Star Trek-style** holodecks may help them escape the isolation of space

(Zolfagharifard, 2014)

Star Trek-like headset lets woman who lost her sight as a child see her husband and baby for the first time

(Best, 2017a)

In these examples, the words highlighted in bold emphasise the comparisons of XR to fiction. As real XR was only just emerging into the consumer market during the period studied, readers may have been unaware of what XR is. However, they are more likely to be familiar with their fictional representations. Therefore, the use of fiction analogies portrays XR as similar to how the technology appears in fiction. Fictional versions of XR are much more advanced than the real products, though this is not noted by these articles. Because of this, associating fictional XR with real devices portrays current day XR as Advanced and High-Quality. Furthermore, this frame is particularly strong in these examples since it appears in the most salient part of the news article – the headline (Pan and Kosicki, 1993). In other words, the rhetorical framing device of fiction analogies is complemented by a technical framing device: the decision of the journalists to include these references in the headline. Since "frames work by connecting the mental dots for the public" (Nisbet, 2010: 47), referencing fiction that audiences may be familiar with could act as a powerful framing device, particularly since it has been appropriated in article headlines.

Still, it is possible that readers may not be aware of these fictional texts. To compensate for this, news articles sometimes used an additional technical framing device to enhance the salience of the Advanced and High-Quality frame: images. One example of this appeared in a *MailOnline* article about a smartphone-based VR headset called Pinć. This device included finger rings that allowed the user to interact with the virtual environment using their hands. One paragraph of the article stated: "a user can make hand gestures to control on-screen objects, in a similar way to the gloves used by Tom Cruise in 2002 sci-fi film Minority Report" (Woollaston, 2014). Again, comparing the product to a well-known film makes an association between real XR and the advanced technology presented in fiction. However, for those unfamiliar with *Minority Report*, a picture of actor Tom Cruise using the gloves was also included as a visual comparison. Therefore, even those unaware of *Minority Report* could make the association between the fictional technology and the real product. This increases the salience of the Advanced and High-Quality frame since these references could be more widely understood by the audience. Coleman argues that "images exert a more powerful influence on memory and perceptions than text" (2010: 243). Thus, the use of imagery to depict the Advanced and High-Quality frame further demonstrates the prominence of this frame.

As well as simply associating fictional XR with actual XR, news articles also argued that the current generation of products represent fiction becoming a reality. For instance, the headline of a *MailOnline* article stated: " 'Holodeck' becomes a reality" (O'Callaghan, 2014) and another *MailOnline* article described Google Glass as "[s]traight out of science-fiction predictions of what future homes will be like" (Nye, 2013). Similarly, the introductory sentence of a *Guardian* article noted: "It might look like a scene from *Minority Report*, but Constantinos Miltiadis's hi-tech gear is science fact, not fiction" (Davis, 2015). As mentioned above, the XR technology seen in fiction is advanced and futuristic. Therefore, suggesting these technologies from fiction are becoming real implies that the actual devices are similar to their fictional counterparts, making them appear advanced. Entman states that the "frames that employ more culturally resonant terms have the greatest potential for influence" (2003: 417). Additionally, Chan points out that popular forms of entertainment (such as those mentioned here) can reach millions of viewers and are thus "important cultural resources" (Chan, 2014: 105). This suggests that associating XR with fiction could be a particularly powerful framing device in constructing the Advanced and High-Quality frame.

Moreover, sources have also used as a technical framing device to give this claim credibility. In particular, the *Guardian* quoted Mark Zuckerberg, stating: "In just a few years, VR has gone from being this science fiction dream to an awesome reality" (Day, 2015). As well as highlighting fiction has become fact, Zuckerberg presents this as something very positive by using the words "dream" and "awesome". Coleman and Ross state that source choice affects a news story's "shape and orientation, casually but irrevocably promoting a particular perspective which goes unchallenged" (2010: 49). The use of Zuckerberg as a source shows that, within the social institutions factor of the hierarchy of influences model (Shoemaker and Reese, 2014), an XR company owner has acted as a frame advocate and played a role in presenting XR as Advanced and High-Quality. This coincides with the findings regarding the Transcendent and Revolutionary and Transformative frames.

The idea that fiction is becoming real also appeared in marketing of Oculus Rift, Gear VR and HoloLens to construct the Advanced and High-Quality frame. For instance, a Facebook post advertising Gear VR stated: "With the passing of time, a lot of what we used to consider science fiction has become reality" (Oculus, 2015). The post invited the reader to follow a link to learn more about how Gear VR was made. This was accompanied by images showing various stages of Gear VR development. Combining this sentence about science fiction with the link and images implies that the creation of Gear VR is one example of fictional technology becoming real. In a similar way, the HoloLens "Possibilities" video advert included a developer of the device stating that what the HoloLens does was once "science fiction and now we're bringing it into science fact" (Microsoft, 2015a). This was repeated on their website, which included "science fiction becomes science fact" as a section heading (Microsoft, 2015c). These examples show that the same frame (Advanced and High-Quality) and framing device (fiction becoming real) have been used in both the news and marketing discourses, demonstrating another similarity between them. As mentioned above, the repetition of frames in different media enhances their persuasive power (Van Gorp, 2007). It appears the Advanced and High-Quality frame is no exception to this, with both news and marketing discourses reinforcing the frame.

The use of fiction references to frame XR as Advanced and High-Quality is surprising considering the often dystopic visions of VR that appear in fiction (Ariel, 2017; Bailenson, 2018; Steinicke, 2016). These connections could have easily been utilised to present real world XR in a negative light and perhaps even to produce moral panic style discourse. Certainly, there were some instances of fiction being used in this way. For example, one *MailOnline* article explained that HoloLens "has been likened to a plot device in an episode of hit Netflix show Black Mirror in which humans are implanted with a gadget that records all that they do, say and hear" (Ellery, 2017). This highlights the privacy concerns surrounding AR mentioned in Chapter 2. However, this statement appeared in the very last paragraph of the article, with the remainder extolling the benefits of HoloLens to help find lost objects, including a quote from the Alzheimer's Society emphasising the benefit to dementia sufferers. Considering that several references to fiction appear in article headlines as a way to represent XR as Advanced and High-Quality, this shows that journalists have favoured positive framing of the technology even when there are clear links to potential negatives.

Although fiction references were a major part of constructing the Advanced and High-Quality frame, this was not the only way the frame could be observed in the news articles. In addition to fiction references, the inclusion and description of product specifications contributed to framing XR as Advanced and High-Quality. An example of this can be seen in an article from the *Guardian* about HTC Vive, which described the headset as a "powerful new VR headset [...] featuring two 1200 × 1080 displays, a smooth 90-frames-per-second refresh rate and a bunch of motion tracking technologies" (Stuart, 2015). Since the *Guardian* is a generalist news outlet (rather than a technology news outlet), a large portion of its audience may not understand what these specifications mean. Indeed, Crow and Stevens note that the use of jargon can reduce the persuasiveness of a message (2012: 112). To avoid this issue, the journalist used a rhetorical framing device in the way that they have described these specifications to attest that the figures correlate to a high-quality experience. The descriptor "powerful" was used first, implying the headset has been built with advanced components capable of providing a high-quality experience. Additionally, the refresh rate was described as "smooth", assuring readers unfamiliar with the terminology that this number equates to a seamless XR experience. Further to this, the use of "bunch" in relation to the device's motion tracking creates the impression that the device is well-equipped to process movement, making it capable of providing a high-quality experience. Here, the typical routine practice (factor four in the hierarchy of influences model (Shoemaker and Reese, 2014)) for journalists to make complex issues understandable to a wide audience (Carlson, 2017) has impacted the construction of the Advanced and High-Quality frame.

Moreover, the same article used an additional framing device in the form of quotations to enhance the Advanced and High-Quality frame. Developers of VR content were quoted positively evaluating the specifications of HTC Vive. One developer claimed: "The specs sound pretty solid [...] and the screen resolution seems good. The tracking volume of 15 ft sounds excellent" (Stuart, 2015). As in the previous example, readers may be unfamiliar with terms such as "specs", "resolution" and "tracking volume". However, the developer's quote includes the descriptors "pretty solid", "good" and "excellent" to clarify that these details are positive. Furthermore, the same developer also noted that HTC Vive "now puts two HMDs on the market aimed at the highest possible consumer VR experience" (Stuart, 2015). The use of the superlative "highest possible" argues that there could be nothing better than this device, placing strong emphasis on the Advanced and High-Quality frame. Additionally, how sources are labelled signifies their level of authority (Bell, 1991: 193). The credibility of these statements is supported by the apparent expertise of the source in the area since the journalist labelled him as a VR developer, suggesting he is knowledgeable about the industry and the quality of VR products. However, developers of VR content are invested in the success of the technology because they need their content to sell. Therefore, it is not surprising that this developer speaks very positively about HTC Vive. The journalist's choice to include the developer's words in the news article indicates that advocates of the Advanced and High-Quality frame have impacted the frame-building process within the social institutions factor (Shoemaker and Reese, 2014).

Furthermore, XR promotional materials also used technical jargon alongside descriptive modifiers to frame the technology as Advanced and High-Quality. This approach appeared in marketing of both HoloLens and Gear VR. Regarding HoloLens, a page of the device's website stated it has "[s]pecialized components – like multiple sensors, advanced optics, and a custom holographic processing unit" (Microsoft, 2016a). As above, for audiences who may not understand these technical aspects, the use of the modifiers "specialised", "multiple", "advanced" and "custom" are added to highlight that the product is advanced. Similarly, the Gear VR website included the following description of the headset:

> It's a clearly superior virtual reality experience with the wide 101 [degree] field of view through the large lens and the smooth and precise head tracking via the built-in gyro sensor and accelerometer.
>
> (Samsung, 2017)

Again, although technical details are mentioned, they are combined with descriptions such as "clearly superior" and "smooth and precise" to illustrate that these features create a high-quality experience. Steinicke argues that, for a compelling VR experience, "one needs high-quality visual graphics, displayed at interactive frame rates, high resolution, precise and accurate tracking, fast connection, and low end-to-end latency" (2016: 15). Therefore, it is not surprising that XR marketing has promoted these technical aspects as being high-quality. However, what is significant is that the same occurs in the news coverage, thus reinforcing the promotional Advanced and High-Quality frame.

Nevertheless, it is important to acknowledge that some news articles countered the Advanced and High-Quality frame. Such instances centred around the idea that XR requires further development before it can be successful. For example, in a *MailOnline* side-note that was used in 16 articles from March 2014 to January 2016, Oculus Rift was described as "not quite ready for primetime yet" (first appearing in an article by Prigg (2014)). This suggests the device must be developed further before it is ready for a mainstream audience. Importantly, these articles were published in the period before the consumer Oculus Rift headset had been released, meaning it was still in its development phase. When the consumer Oculus Rift *was* released, another *MailOnline* article still highlighted this idea: "Reviewers claim the Facebook-owned device is a 'wonderfully immersive' device, but it still has a way to go" (Woollaston, 2016). Here, although immersion is implied to be high-quality, the product is still said to need improvements in other areas. Therefore, there have been some attempts to counter the Advanced and High-Quality frame, at least in the *MailOnline*. However, this is not to the extent that it would eclipse the multiple framing devices used to portray XR as Advanced and High-Quality discussed above.

These framing strategies differ from previous studies in the way that fiction is used in technology news. For instance, Petersen, Anderson and Allan state that science fiction imagery is often used in news that discusses "the powers and dangers of biotechnology" (2005: 338–339). On the other hand, in this study, fiction has been used in the news coverage of XR to frame it as Advanced and High-Quality, avoiding moral panic style discourse. Within lifestyle journalism about cultural products, previous studies have found that positive evaluations of quality in media have contributed to legitimising these products (Kristensen, Hellman and Riegert, 2019: 259). It stands to reason, then, that positively evaluating the quality of XR could have the same legitimising effect.

Furthermore, while framing XR as advanced could make it appear complex and thus reduce the likelihood of it being adopted (Rogers, 2003), this does not seem to be the case here. Instead, the Advanced and High-Quality frame has emphasised the relative advantage of XR, another of Rogers' (2003) characteristics of successful innovations. Additionally, Kotler et al. state that one of the main goals of marketing "is to attract new customers by promising superior value" (2016: 4). Since the Advanced and High-Quality frame highlights the superior value of XR, it appears that the news coverage has aided this goal of marketing by applying this frame to XR. While it is not surprising that this frame appeared in the marketing, what *is* significant is that it was also prominent in XR news coverage, indicating a blurring between news and promotional content. This news appears to have been marketised (Fairclough, 1993), supporting Chyi and Lee's argument that, in technology news, "the boundary between news and promotional content is tenuous at best" (2018: 585). The positive, potentially promotional, tone of the Advanced and High-Quality frame is yet another example of frames being shared between the news and marketing that could create a positive view of XR and thus support its adoption.

The dominance of these positive frames means that, instead of prioritising the interests of the general public by paying attention to both the benefits and risks surrounding XR, the news prioritises the commercial interests of XR companies. This compromises the fourth estate role of journalism. Chapter 3 discussed the presence of native advertising within the news articles, as well as the fact some articles were written by creators of XR content. On the social institutions level (Shoemaker and Reese, 2014), both of these points indicate that the news outlets have relationships with XR companies that they are invested in maintaining for their own commercial gain. In other words, to continue to make money within the capitalist social system, news organisations want to maintain these relationships. This could be one reason why the outlets prioritise these positive frames and pay very little attention to critical viewpoints surrounding XR. Further insight will be gleaned in the remaining chapters of this book.

References

Apple (2018) *Apple Reports First Quarter Results* [Press Release]. 1 February. Available at: www.apple.com/uk/newsroom/2018/02/apple-reports-first-quarter-results/ (Accessed: 22 April 2020).

Ariel, G. (2017) *Augmenting Alice*. Amsterdam: BIS Publishers.

Arthur, C. (2015) 'The return of virtual reality: 'this is as big an opportunity as the internet", *The Guardian*, 28 May. Available at: www.theguardian.com/technology/2015/may/28/jonathan-waldern-return-virtual-reality-as-big-an-opportunity-as-internet (Accessed: 20 December 2018).

Associated Press (2015) 'Virtual reality is here: Samsung's $99 Gear VR launches (but you might need a new phone to use it)', *MailOnline*, 20 November. Available at: www.dailymail.co.uk/sciencetech/article-3327249/Review-Samsungs-Gear-VR-shows-promise-VR-_-today.html (Accessed: 13 February 2019).

Bailenson, J. (2018) *Experience on Demand*. New York: W.W. Norton & Company.

Bauer, M.W., Howard, S., Romo Ramos, Y.J., Massarani, L. and Amorim, L. (2013) *Global Science Journalism Report: Working Conditions & Practices, Professional Ethos and Future Expectations*. London: Science and Development Network. Available at: http://eprints.lse.ac.uk/48051/ (Accessed: 23 April 2021).

Bednarek, M. and Caple, H. (2012) *News Discourse*. London: Continuum International Publishing Group.

Bell, A. (1991) *The Language of News Media*. Oxford: Blackwell.

Best, S. (2017a) 'Star Trek-like headset lets woman who lost her sight as a child see her husband and baby for the first time', *MailOnline*, 20 February. Available at: www.dailymail.co.uk/sciencetech/article-4242456/Star-Trek-like-glasses-allow-blind-people-see.html (Accessed: 11 February 2019).

Best, S. (2017b) 'Apple's Tim Cook predicts augmented reality will be bigger than VR because it doesn't isolate people in their own worlds', *MailOnline*, 12 October. Available at: www.dailymail.co.uk/sciencetech/article-4973026/Tim-Cook-SLAMS-virtual-reality-new-interview.html (Accessed: 7 February 2019).

Bostrom, N. (2007) 'Technological Revolutions: Ethics and Policy in the Dark', in Cameron, N.M. and Mitchell, M.E. (eds.) *Nanoscale: Issues and Perspectives for the Nano Century*. New Jersey: Wiley, pp. 129–152.

Burrows, T. (2015) 'Generals will be able to direct battles using new minority report-style technology including 3D goggles and even virtual reality contact lenses', *MailOnline*, 10 May. Available at: www.dailymail.co.uk/news/article-3075485/Headsets-aid-military-commanders.html (Accessed: 13 February 2019).

Carlson, M. (2017) *Journalistic Authority: Legitimating News in the Digital Era*. New York: Columbia University Press.

Carty, S. (2015) "It was a bit like the Matrix': FIFO father becomes world's first man to experience son's birth from 4000km away after breakthrough in virtual technology', *MailOnline*, 14 March. Available at: www.dailymail.co.uk/news/article-2994568/It-bit-like-Matrix-FIFO-father-world-s-man-experience-son-s-birth-4000km-away-breakthrough-virtual-technology.html (Accessed: 13 February 2019).

Chan, M. (2014) *Virtual Reality: Representations in Contemporary Media*. London: Bloomsbury.

Chyi, H.I. and Lee, A.M. (2018) 'Commercialization of Technology News', *Journalism Practice*, 12(5), pp. 585–604. doi:10.1080/17512786.2017.1333447.

Cohen, S. (2002) *Folk Devils and Moral Panics*. 3rd edn. Oxon: Routledge.

Coleman, R. (2010) 'Framing the Pictures in Our Heads: Exploring the Framing and Agenda-Setting Effects of Visual Images', in D'Angelo, P. and Kuypers, J.A. (eds.) *Doing News Framing Analysis*. Oxon: Routledge, pp. 233–261.

Coleman, S. and Ross, K. (2010) *The Media and the Public*. West Sussex: Wiley-Blackwell.

Cooper, R.G. (1979) 'The Dimensions of Industrial New Product Success and Failure', *Journal of Marketing*, 43(3), pp. 93–103. doi:10.2307/1250151.

Costandi, M. (2012) 'Inception helmet creates alternative reality', *The Guardian*, 26 August. Available at: www.theguardian.com/science/neurophilosophy/2012/aug/26/inception-helmet-alternative-reality (Accessed: 21 December 2018).

Crow, D.A. and Stevens, J.R. (2012) 'Framing Science: The Influence of Expertise and Jargon in Media Coverage', *Proceedings of the 2012 Summer Symposium on Science Communication: Between Scientists & Citizens*, Iowa, USA, 1–2 June, pp. 109–119. doi:10.31274/sciencecommunication-180809-61.

Davis, N. (2015) 'Project Anywhere: digital route to an out-of-body experience', *The Guardian*, 7 January. Available at: www.theguardian.com/technology/2015/jan/07/project-anywhere-digital-route-to-an-out-of-body-experience (Accessed: 19 December 2018).

Day, E. (2015) 'Virtual reality? Not for me. Then I turn into Wonder Woman and fly over New York', *The Guardian*, 11 October. Available at: www.theguardian.com/technology/2015/oct/11/virtual-reality-oculus-rift-stanford-silicon-valley-facebook (Accessed: 21 December 2018).

Desrosiers, M. (2012) 'Reframing Frame Analysis: Key Contributions to Conflict Studies', *Ethnopolitics*, 11(1), pp. 1–23. doi:10.1080/17449057.2011.567840.

Dickerson, D.L. (2001) 'Framing "Political Correctness": *The New York Times'* Tale of Two Professors', in Reese, S.D., Gandy, O.H. and Grant, A.E. (eds.) *Framing Public Life*. London: Lawrence Erlbaum Associates, pp. 163–174.

Duffy, M.J. and Freeman, C.P. (2011) 'Unnamed Sources: A Utilitarian Exploration of Their Justification and Guidelines for Limited Use', *Journal of Mass Media Ethics*, 26(4), pp. 297–315. doi:10.1080/08900523.2011.606006.

Ellery, B. (2017) 'What a find! The virtual reality glasses that remember where you've left your car keys: new technology recognises objects and 'tracks' their whereabouts for the wearer', *MailOnline*, 8 January. Available at: www.dailymail.co.uk/sciencetech/article-4098364/Virtualy-reality-glasses-remember-ve-left-important-items.html (Accessed: 11 February 2019).

Entman, R.M. (1993) 'Framing: Towards Clarification of a Fractured Paradigm', *Journal of Communication*, 43(4), pp. 51–58.

Entman, R.M. (2003) 'Cascading Activation: Contesting the White House's Frame After 9/11', *Political Communication*, 20, pp. 415–432. doi:10.1080/10584600390244176.

Facebook (2014) *Facebook to Acquire Oculus*. 25 March [Press Release]. Available at: https://newsroom.fb.com/news/2014/03/facebook-to-acquire-oculus/ (Accessed: 31 October 2019).

Facebook (2015) *Introducing the Oculus Rift.* 11 June [Press Release]. Available at: https://newsroom.fb.com/news/2015/06/introducing-the-oculus-rift/ (Accessed: 31 October 2019).

Fairclough, N. (1993) 'Critical Discourse Analysis and the Marketization of Public Discourse: The Universities', *Discourse & Society*, 4(2), pp. 133–168.

Faithfull, M. (2022) 'Luxury soars but apple named best global brand for tenth year running', *Forbes*, 3 November. Available at: www.forbes.com/sites/markfaithfull/2022/11/03/luxury-soars-but-apple-named-best-global-brand-for-tenth-year-running/ (Accessed: 2 June 2023).

Flight, R.L., Allaway, A.W., Kim, W. and D'Souza, G. (2011) 'A Study of Perceived Innovation Characteristics Across Cultures and Stages of Diffusion', *Journal of Marketing Theory and Practice*, 19(1), pp. 109–126. doi:10.2753/MTP1069-6679190107.

Gandy, O.H. (1982) *Beyond Agenda Setting: Information Subsidies and Public Policy.* Norwood, NJ: Ablex Publishing Corporation.

Gibbs, S. (2014) 'Sony's Project Morpheus brings virtual reality to mainstream console gaming', *The Guardian*, 12 May. Available at: www.theguardian.com/technology/2014/may/12/sonys-project-morpheus-virtual-reality-console-gaming (Accessed: 19 December 2018).

Gibbs, S. (2016) 'Sky sets up in-house studio for virtual reality content', *The Guardian*, 17 March. Available at: www.theguardian.com/technology/2016/mar/17/sky-sets-up-in-house-studio-virtual-reality-content (Accessed: 19 December 2018).

Goggin, G. (2006) *Cell Phone Culture: Mobile Technology in Everyday Life.* Oxon: Routledge.

Griffiths, S. and Prigg, M. (2015) 'Magic Leap set to revolutionise every aspect of daily life: patent of 'secret' augmented reality headset reveals uses in shops, hospitals and homes', *MailOnline*, 19 January. Available at: www.dailymail.co.uk/sciencetech/article-2916696/Magic-Leap-set-revolutionise-aspect-daily-life-Patent-secret-augmented-reality-headset-reveals-uses-shops-hospitals-homes.html (Accessed: 14 February 2019).

Hecht, R.D., Martin, F., Donnelly, T., Larson, M. and Sweetser, K.D. (2017) 'Will You Run It? A Gatekeeping Experiment Examining Credibility, Branding, and Affiliation within Information', *Public Relations Review*, 43(4), pp. 738–749. doi:10.1016/j.pubrev.2017.07.006.

Judge, J. (2016) 'Facebook's virtual reality just attempts what artists have been doing forever', *The Guardian*, 28 July. Available at: www.theguardian.com/technology/2016/jul/28/facebook-virtual-reality-oculus-rift-mark-zuckerberg-art (Accessed: 13 December 2018).

Kelly, J.P. (2009) 'Not so Revolutionary after All: The Role of Reinforcing Frames in US Magazine Discourse about Microcomputers', *New Media & Society*, 11(1–2), pp. 31–52. doi:10.1177/1461444808100159.

Kickstarter (2012) *Oculus Rift: Step into the Game.* Available at: https://web.archive.org/web/20120801212942/http:/www.kickstarter.com/projects/1523379957/oculus-rift-step-into-the-game (Accessed: 6 September 2019).

Kotler, P., Armstrong, G., Harris, L.C. and Piercy, N. (2016) *Principles of Marketing.* 7th European Edition edn. Harlow: Pearson Education Limited.

Kristensen, N.N., Hellman, H. and Riegert, K. (2019) 'Cultural Mediators Seduced by *Mad Men*: How Cultural Journalists Legitimized a Quality TV Series in the

Nordic Region', *Television & New Media*, 20(3), pp. 257–274. doi:10.1177/ 1527476417743574.

Krumsvik, A.H., Milan, S., Bhroin, N.N. and Storsul, T. (2019) 'Making (Sense of) Media Innovations', in Deuze, M. and Prenger, M. (eds.) *Making Media: Production, Practices, and Professions*. Amsterdam: Amsterdam University Press, pp. 193–205.

Lecheler, S. and de Vreese, C.H. (2019) *News Framing Effects*. Oxon: Routledge.

Linström, M. and Marais, W. (2012) 'Qualitative News Frame Analysis: A Methodology', *Communitas*, 17, pp. 21–28.

Macdonald, C. (2016) 'Star Wars-style moving holograms are here: Microsoft shows how HoloLens can bring distant family members into your home', *MailOnline*, 28 March. Available at: www.dailymail.co.uk/sciencetech/article-3513062/Star-Wars-style-moving-holograms-Microsoft-shows-HoloLens-bring-distant-family-memb ers-home.html (Accessed: 7 December 2018).

Magic Leap (2014) *Magic Leap Raises more Than $50 Million*. 5 February [Press Release]. Available at: www.magicleap.com/en-us/news/press-release/magic-leap-raises-50million (Accessed: 16 September 2019).

Magic Leap (2015) *Developers*. Available at: https://web.archive.org/web/2015022 5015415/https:/www.magicleap.com/#/developers (Accessed: 16 September 2019).

Magic Leap (2016a) *Magic Leap Announces Expanded Role for Weta Workshop's Sir Richard Taylor*. 2 February [Press Release]. Available at: www.magicleap. com/en-us/news/press-release/sir-richard-taylor-joins-magic-leap (Accessed: 16 September 2019).

Magic Leap (2016b) '47 years ago, one giant leap inspired our name. Today, we still believe in the unimpossible' [Facebook] 20 July. Available at: www.facebook.com/ magicleap (Accessed: 16 September 2019).

Magic Leap (2017) 'This is the story of an art project, turned tech startup, turned global company building the future of computing' [Twitter] 5 October. Available at: https://twitter.com/magicleap/status/915962784537628672 (Accessed: 12 September 2019).

Markey, P.M. and Ferguson, C.J. (2017) 'Teaching Us to Fear', *American Journal of Play*, 10(1), pp. 99–115.

Microsoft (2015a) *Microsoft HoloLens*. Available at: https://web.archive.org/web/ 20150124143645/http:/www.microsoft.com/microsoft-hololens/en-us (Accessed: 3 September 2019).

Microsoft (2015b) *Microsoft HoloLens – Transform Your World with Holograms*. 21 January. Available at: www.youtube.com/watch?v=aThCr0PsyuA (Accessed: 2 October 2019).

Microsoft (2015c) *Hardware*. Available at: https://web.archive.org/web/20150728065 657/http:/www.microsoft.com/microsoft-hololens/en-us/hardware (Accessed: 13 September 2019).

Microsoft (2015d) *Developers*. Available at: https://web.archive.org/web/201 51209171938/http:/www.microsoft.com/microsoft-hololens/en-us/developers (Accessed: 12 September 2019).

Microsoft (2016a) *Development Edition*. Available at: https://web.archive.org/web/ 20160304165140/https:/www.microsoft.com/microsoft-hololens/en-us/developm ent-edition (Accessed: 13 September 2019).

Microsoft (2016b) *Commercial*. Available at: https://web.archive.org/web/201 60315185619/http:/www.microsoft.com/microsoft-hololens/en-us/commercial (Accessed: 3 September 2019).

Microsoft (2016c) *Microsoft HoloLens.* Available at: https://web.archive.org/web/201 61031095718/http://www.microsoft.com/microsoft-hololens/en-us (Accessed: 12 September 2019).

Microsoft (2016d) *Developers.* Available at: https://web.archive.org/web/201 61115183651/https:/www.microsoft.com/microsoft-hololens/en-us/developers (Accessed: 13 September 2019).

Microsoft (2016e) *Commercial.* Available at: https://web.archive.org/web/201 60315185619/http:/www.microsoft.com/microsoft-hololens/en-us/commercial (Accessed: 3 September 2019).

Microsoft HoloLens (2016) 'See detail like never before when you build in 3D with the Development Edition' [Twitter] 22 June. Available at: https://twitter.com/HoloL ens/status/745651321638895616 (Accessed: 12 September 2019).

Nisbet, M.C. (2010) 'Knowledge into Action: Framing the Debates Over Climate Change and Poverty', in D'Angelo, P. and Kuypers, J.A. (eds.) *Doing News Framing Analysis.* Oxon: Routledge, pp. 43–83.

Nordfors, D. (2009) 'Innovation Journalism, Attention Work and the Innovation Economy', *Innovation Journalism*, 6(1), pp. 1–46.

Nye, J. (2013) 'Google Glass could allow users to control household appliances like their coffee machine and garage doors', *MailOnline*, 22 March. Available at: www.dailymail.co.uk/news/article-2297310/Google-Glass-allow-users-control-household-appliances-like-coffee-machine-garage-doors.html (Accessed: 12 December 2018).

O'Callaghan, J. (2014) "Holodeck' becomes a reality: Star Trek-style system uses a wireless Oculus Rift to visit virtual worlds', *MailOnline*, 21 July. Available at: www.dailymail.co.uk/sciencetech/article-2700073/Holodeck-reality-Star-Trek-style-uses-wireless-Oculus-Rift-visit-virtual-worlds.html (Accessed: 14 February 2019).

Oculus (2015) 'What if, sitting on the couch, you can dive into the ocean and explore the underwater world or check out Tony Stark's lab from the Avengers?' [Facebook] 20 January. Available at: www.facebook.com/Oculusvr (Accessed: 16 September 2019).

Oculus (2017a) 'Take a step into the future, 2049 to be exact' [Facebook] 29 October. Available at: www.facebook.com/OculusGB (Accessed: 3 April 2020).

Oculus (2017b) 'Leave the mundane behind' [Facebook] 25 May. Available at: www.facebook.com/OculusGB (Accessed: 3 April 2020).

Oculus VR (2012) *New Virtual Reality Gaming Headset from Oculus Gets Kickstarted.* 1 August [Press Release]. Available at: https://web.archive.org/web/20120811020 422/http://oculusvr.com:80/press_release/ (Accessed: 21 December 2018).

Pan, Z. and Kosicki, G.M. (1993) 'Framing Analysis: An Approach to News Discourse', *Political Communication*, 10, pp. 55–75.

Petersen, A., Anderson, A. and Allan, S. (2005) 'Science Fiction/Science Fact: Medical Genetics in News Stories', *New Genetics and Society*, 24(3), pp. 337–353. doi:10.1080/14636770500350088.

Plummer, L. and O'Hare, R. (2016) 'PlayStation VR is now on sale: Sony's virtual reality gaming headset will take on the Oculus Rift and HTC Vive', *MailOnline*, 12 October. Available at: www.dailymail.co.uk/sciencetech/article-3834393/Sony-tapping-virtual-reality-PlayStation-headset.html (Accessed: 11 February 2019).

Poole, S. (2014) 'What does the Oculus Rift backlash tell us? Facebook just isn't cool', *The Guardian*, 27 March. Available at: www.theguardian.com/commentisf ree/2014/mar/27/oculus-rift-facebook-buy-out-kickstarter (Accessed: 21 December 2018).

Prigg, M. (2013) 'How Google Glass projects data directly to your eye: Secrets of how search giant's controversial new gadget revealed', *MailOnline*, 9 April. Available at: www.dailymail.co.uk/sciencetech/article-2306382/How-Glass-works-New-infographic-reveals-secrets-Googles-interactive-eyewear.html (Accessed: 12 December 2018).

Prigg, M. (2014) Facebook buys virtual reality headset firm Oculus for $2bn as Mark Zuckerberg promises to 'change the way we communicate', *MailOnline*, 25 March. Available at: www.dailymail.co.uk/sciencetech/article-2589367/Get-ready-social-platform-Facebook-buys-virtual-reality-firm-Oculus-2bn.html (Accessed: 14 February 2019).

Roderick, I. (2016) *Critical Discourse Studies and Technology*. London: Bloomsbury.

Rogers, E.M. (2003) *Diffusion of Innovations*. 5th edn. New York: Free Press.

Rogers, R. (2013) 'Critical Essay — Old Games, Same Concerns', *Technoculture*, 3. Available at: https://tcjournal.org/drupal/vol3/rogers (Accessed: 10 October 2016).

Samsung (2016) *Gear VR*. Available at: https://web.archive.org/web/20160804072 543/http:/www.samsung.com/global/galaxy/gear-vr/ (Accessed: 29 August 2019).

Samsung (2017) *Gear VR*. Available at: https://web.archive.org/web/20170330014 035/http:/www.samsung.com/global/galaxy/gear-vr/#!/ (Accessed: 29 August 2017).

Samsung US (2015) *Gear VR Demonstration*. 22 January. Available at: https://youtu. be/-gnvQS2xhRg (Accessed: 29 August 2019).

Schäfer, M.S. (2017) 'How Changing Media Structures Are Affecting Science News Coverage', in Jamieson, K.H., Kahan, D. and Scheufele, D. (eds.) *Oxford Handbook on the Science of Science Communication*. New York: Oxford University Press, pp. 51–60.

Sheafer, T., Shenhav, S.R. and Amsalem, E. (2018) 'International Frame Building in Mediated Public Diplomacy', in D'Angelo, P. (ed.) *Doing News Framing Analysis II*. Oxon: Routledge, pp. 249–273.

Shoemaker, P.J. and Reese, S.D. (2014) *Mediating the Message in the 21st Century*. Oxon: Routledge.

Spettigue, S. (2015) "Intuitive controls and gorgeous visuals': Dailymail.com goes 'heads on' with Sony's PlayStation VR as experts predict it could outsell Facebook's Oculus Rift', *MailOnline*, 28 October. Available at: www.dailymail.co.uk/sciencet ech/article-3292877/Dailymail-com-goes-heads-Sony-s-PlayStation-VR-experts-predict-outsell-Facebook-s-Oculus-Rift.html (Accessed: 13 February 2019).

Steinicke, F. (2016) *Being Really Virtual*. Cham, Switzerland: Springer.

Stuart, K. (2015) 'HTC Vive: developers react to Valve's virtual reality headset', *The Guardian*, 2 March. Available at: www.theguardian.com/technology/2015/ mar/02/htc-vive-developers-react-to-valve-virtual-reality-headset (Accessed: 14 December 2018).

Sturgis, I. (2015) 'Virtual reality SKIING! World's first true augmented reality ski goggles let adrenaline junkies create slalom tracks to follow and even video message friends on the slopes', *MailOnline*, 21 January. Available at: www.dailymail.co.uk/ travel/travel_news/article-2918199/Virtual-reality-SKIING-World-s-augmented-reality-ski-goggles-let-adrenaline-junkies-create-slalom-tracks-follow-video-mess age-friends-slopes.html (Accessed: 14 February 2019).

Therrien, C. and Lefebvre, I. (2017) 'Now You're Playing with Adverts: A Repertoire of Frames for the Historical Study of Game Culture through Marketing Discourse', *Kinephanos*, 7(November), pp. 37–73.

van Dijk, T.A. (1988) *News as Discourse*. New Jersey: Lawrence Erlbaum Associates.

Van Gorp, B. (2007) 'The Constructionist Approach to Framing: Bringing Culture Back In', *Journal of Communication*, 57, pp. 60–78. doi:10.1111/j.1460-2466.2006.00329.x.

Van Gorp, B. (2010) 'Strategies to Take Subjectivity Out of Framing Analysis', in D'Angelo, P. and Kuypers, J.A. (eds.) *Doing News Framing Analysis*. Oxon: Routledge, pp. 84–109.

Waugh, R. (2012) 'Google glasses with built-in Terminator-style computer displays 'could be on sale by the year's end at a cost of $250", *MailOnline*, 22 February. Available at: www.dailymail.co.uk/sciencetech/article-2097879/Google-glasses-sale-end-2012-cost-250.html (Accessed: 20 February 2019).

Williams, A. and Clifford, S. (2009) *Mapping the Field: Specialist Science News Journalism in the UK National Media*. The Risk, Science and the Media Research Group, Cardiff University. Available at: http://orca.cf.ac.uk/18447/ (Accessed: 8 April 2021).

Williams, S. (2015) 'Game over for joysticks? Fove virtual reality headset lets players aim and interact with characters using just their EYES', *MailOnline*, 19 May. Available at: www.dailymail.co.uk/sciencetech/article-3087715/Game-joysticks-Fove-virtual-reality-headset-lets-players-aim-interact-characters-using-EYES.html (Accessed: 13 February 2019).

Woollaston, V. (2014) 'Turn your iPhone into a virtual reality headset: £63 Pinć uses optical rings and hand gestures to control the screen', *MailOnline*, 26 November. Available at: www.dailymail.co.uk/sciencetech/article-2850203/Turn-iPhone-virtual-reality-headset-63-Pinc-uses-optical-rings-hand-gestures-control-screen.html (Accessed: 14 February 2019).

Woollaston, V. (2016) 'Oculus Rift begins shipping: customers who booked early are now starting to receive their $599 virtual reality headsets', *MailOnline*, 29 March. Available at: www.dailymail.co.uk/sciencetech/article-3512974/Oculus-Rift-begins-shipping-reviews-suggest-waiting-OK.html (Accessed: 12 February 2019).

Yadron, D. (2016) 'We've seen Magic Leap's device of the future, and it looks like Merlin's skull cap', *The Guardian*, 8 June. Available at: www.theguardian.com/technology/2016/jun/07/magic-leap-headset-design-patent-virtual-reality (Accessed: 21 December 2018).

Zolfagharifard, E. (2014) 'Could virtual reality prevent depression in ASTRONAUTS? Star Trek-style holodecks may help them escape the isolation of space', *MailOnline*, 15 October. Available at: www.dailymail.co.uk/sciencetech/article-2793768/could-virtual-reality-prevent-depression-astronauts-star-trek-style-holodecks-help-escape-isolation-space.html (Accessed: 14 February 2019).

6 Framing a Satisfying User Experience

Rogers notes that there are usually two components to a technology: "(1) a *hardware* aspect, consisting of the tool that embodies the technology as a material or physical object, and (2) a *software* aspect, consisting of the information base for the tool" (2003: 13, original emphasis). The third frame category in the model for analysing media coverage of emerging technologies, User Experience, focuses on those two areas. This category invites the researcher to consider how the act of using the technology is portrayed, both in terms of its hardware and its software. To identify frames in this category, the following questions could be asked:

- How is the hardware and/or software presented?
- What uses of the technology are given the most/least attention?
- What aspects of the user experience are said to be strong/weak?

In the context of XR, hardware refers to the devices (typically headsets) used to display virtual worlds or objects, while software refers to any applications that can be used with the devices. Three frames were identified in the User Experience category in relation to XR, which are the focus of this chapter. Two of the frames (Easy to Use and Comfortable) relate to the hardware aspect of XR, whereas one frame (Social) refers to the software aspect of XR. These frames appeared in the news and marketing of XR; thus, each section of this chapter considers the framing devices used in both samples and how these relate to each other.

Social

Since VR literally isolates the user from the physical environment by requiring them to wear a headset that blocks out the real world (Brigham, 2017), it is perhaps surprising that one of the frames that emerged was Social. This section examines the construction of the Social frame in the news articles and how it relates to the marketing materials.

DOI: 10.4324/9781003375814-6

Firstly, evidence of the Social frame appearing in the news articles can be seen in the use of certain terms. Words in the "social" category appeared in 12 percent of articles overall. While this is by no means the majority of articles, when comparing this to words describing XR as isolating, the difference is stark. The search terms *isolat** (e.g. isolating, isolation) and *solitary* appeared 39 times combined, whereas words in the "social" category were used 235 times. These figures show that words referring to XR as social were used over six times more than those describing XR as isolating. Even the word *social* alone appeared 65 times; 1.7 times more than *isolat** and *solitary* combined. Therefore, it is clear that the news articles were more likely to frame XR as Social than isolating. This shows that word choice has acted as a framing device in the construction of the Social frame.

However, there were some noticeable differences between the use of these words in the individual news outlets. The *Guardian* was most likely to use words from the "social" category, with 17 percent of articles from this outlet mentioning such words at least once. With a slightly lower portion, 11 percent of *MailOnline* articles used words in the "social" category. The largest difference can be seen in the *Sun* which only used "social" words in 3 percent of its articles. This suggests that the *Sun* rarely used the Social frame. Still, all three publications used *isolate** and *solitary* less than they used words in the "social" category. In this way, although the media organisation (factor three in the hierarchy of influences model (Shoemaker and Reese, 2014)) reporting on XR appears to have influenced the strength of the Social frame, this factor has not affected the news discourse to the extent that contrasting frames are used in different news outlets.

Comparing the use of words in the "social" category based on XR type shows further evidence that the physically isolating nature of VR has not lessened the strength of the Social frame. Terms in the "social" category appeared in 12 percent of articles about VR, compared to 10 percent of AR/MR articles. These figures suggest that, despite VR headsets being more physically isolating than AR/MR products because they cover the users view of the real world, this has not prevented journalists from framing VR as Social. Certainly, when the news coverage focused on AR/MR in 2012 and 2013, words referring to the Social frame rarely appeared, if at all. On the other hand, the percentage of articles using words in this category peaked in 2014 (20 percent) when coverage shifted to focus on VR with Mark Zuckerberg's acquisition of Oculus. Considering Zuckerberg's link with the major social network Facebook, it appears that his involvement in XR contributed to framing it as Social. This suggests that frame advocates within the social institutions factor of the hierarchy of influences (Shoemaker and Reese, 2014) have had an impact on the building of the Social frame.

Indeed, as has been the case for other frames, sourcing decisions have been used as a technical framing device to construct the Social frame. In the following example, a quotation from Zuckerberg was used that originated in a press release announcing Facebook's acquisition of Oculus. In the press release, Zuckerberg stated that "Oculus has the chance to create the most social platform ever" (Facebook, 2014). Regarding the news sample, this quote appeared in two articles from the *Guardian* and seven *MailOnline* articles. Similarly, the same press release also quoted the then CEO of Oculus, Brendan Iribe, stating: "We believe virtual reality will be heavily defined by social experiences that connect people in magical, new ways" (Facebook, 2014). Here, Iribe stresses the importance of social VR experiences and claims it will enable "magical" connection, applying a strongly positive connotation to VR being social. Just as with Zuckerberg's words, Iribe's message was transferred to the news sample, with the quotation appearing in one *Guardian* article and five *MailOnline* articles. These examples highlight several important points. Firstly, the choice of journalists to use quotations from a press release brings this information into the public domain (Nordfors, 2009), thus allowing the source to reach the general public. Secondly, the routine practice (hierarchy of influences factor four (Shoemaker and Reese, 2014)) for the *MailOnline* to copy and paste parts of its articles has emphasised this frame that is advocated by XR company owners. Thirdly, the combination of points one and two mean that these frame advocates within the social institutions factor have played a substantial role in the creation of the Social frame. As mentioned in Chapter 5 regarding the Revolutionary and Transformative frame, the repetition of quotes is perhaps a result of the increasing commercial pressures on journalists. This means the social systems factor (Shoemaker and Reese, 2014) has played a role in the frame-building process in this way.

As well as sourcing techniques, news articles used the concept of telepresence as an additional framing device in the creation of the Social frame. Steuer defines telepresence as "the extent to which one feels present in the mediated environment, rather than in the immediate physical environment" (1992: 76). In an opinion article from the *Guardian*, a journalist explained her encounter with a 15-year-old boy using the VR application AltspaceVR. She summed up her experience in the final paragraph of the article: "when I stood next to him, I felt aware of our closeness, despite the 1,300 miles separating our physical bodies" (Evans, 2016). Here, the writer argues that VR allowed her to feel close to someone who was physically very far away, alluding to the idea of telepresence. In another article from the *Guardian*, AR was said to allow users to "meet a friend for coffee at their kitchen table, even if the friend is on another continent" (Yadron, 2016). This sentence suggests users will be able to have authentic social experiences regardless of geographical distance. Similarly, a *MailOnline* article about HoloLens explained how the MR Skype application works: "Developers will be able to see the other person's hologram during the two-way Skype call and interact with them as if they were sitting next to them on the couch" (Liberatore, 2016). Though this sentence does not mention the device transcending physical space, it still implies the experience will be natural and realistic, suggesting it will feel as social as a physical meeting. This, again, alludes to telepresence to create the Social frame.

Furthermore, telepresence was also referred to in Gear VR marketing to frame the product as Social. Referring to sharing New Year's Eve celebrations, the Samsung Mobile Twitter account posted a video with the message: "Celebrate together from a thousand miles away" (Samsung Mobile, 2016). This suggests users will feel as if they are in each other's presence when experiencing VR, even if they are physically far apart. Additionally, the Gear VR website included the following description about the Oculus Rooms and Parties application: "Whether you and your friends are worlds apart or practically next door neighbors, Oculus Rooms and Parties are a convenient and fun way to spend time together" (Samsung, 2017). Again, this sentence highlights that users can "spend time together" regardless of physical distance. These results demonstrate that referencing the concept of telepresence has been used as a framing device in both the news and marketing samples. Van Gorp (2007: 69) argues that the repetition of the same frame and framing devices makes it difficult to refute a frame. Therefore, for this frame *and* framing device to be shared between the two samples makes the Social frame particularly strong.

Aside from this direct similarity, the Social frame was depicted in the marketing of AR and MR products by emphasising their value for collaborative working. For instance, the HoloLens website stated that one of the main features of the device is that it allows "[n]ew ways to collaborate and explore" (Microsoft, 2015a), which includes being able to "[s]ee holograms from your colleague's perspective if he's in the next room or on the other side of the world" (Microsoft, 2015a). This not only presents HoloLens as allowing social collaboration, but highlights that it can also transcend geographical boundaries. Similarly, after Google Glass was rebranded as an enterprise product, the website stated: "Glass can connect you with coworkers in an instant [...] Invite others to 'see what you see' through a live video stream so you can collaborate and troubleshoot in real-time" (Google, 2017). Interacting socially to collaborate in a workplace environment is represented as easy, instant and useful, portraying the device as able to enhance social interaction in the workplace. Finally, Magic Leap marketing took a more whimsical approach to highlighting collaboration. Its website stated that the device allows users to "[c]onnect in physical space with others, digitally. [...] Call it collaboration for another dimension" (Magic Leap, 2017). These examples demonstrate that an exemplar in the form of collaborative working has been used as a framing device to create the Social frame in XR marketing. Moreover, each of these excerpts highlight further evidence of the Social frame being shared between the two samples and thus reinforcing their supposed facticity (Van Gorp, 2007).

However, it is important to examine the few instances that the news sample attempted to counter the Social frame. Unsurprisingly, based on the above discussion, this centred around VR and the fact that its users must wear a headset that replaces their view of the real world with a virtual one (Brigham, 2017). Firstly, the *Guardian* criticised Zuckerberg's social vision for VR in the following article, stating: "the current version of software being created for these headsets is focused on solo experiences while wearing a device that isolates you from the people around you" (Dredge, 2016a). In a similar vein, another article written by the same author in the *Guardian* noted: "There's also the question of isolation, especially when VR involves shutting your-self off from the world around you by wearing a headset" (Dredge, 2016b). Further to this, the same article described a promotional image for Oculus' new social application as "a rather chilling vision of how we might watch TV together in the future" (Dredge, 2016b). It continued on to state: "There's an argument [...] that virtual reality is a next level of physical isolation" (Dredge, 2016b). In each of these cases, the way that a VR headset physic-ally isolates the user has been highlighted as a reason that the technology is isolating. In the final example, instead of suggesting VR can allow people to become more connected by having social experiences that feel more realistic (as above), it is implied that VR could make people even less social. This directly counters the framing device used to present the Social frame in the instances above.

Additionally, whereas quotes from Oculus' owners were used as framing devices for the Social frame, the voice of Apple CEO Tim Cook was used to counter this frame. The headline of a *MailOnline* article read: "Apple's Tim Cook predicts augmented reality will be bigger than VR because it doesn't isolate people in their own worlds" (Best, 2017). By claiming AR contrasts with VR in that it is not isolating, VR itself is shown to be isolating. This sen-timent is given significant prominence since it appears in the headline of the article (Pan and Kosicki, 1993). Moreover, the news item clarified this point in the third paragraph, which stated: "He [Cook] describes VR as isolating" (Best, 2017). As mentioned in previous discussions about the use of Cook's voice (regarding the Transcendent and Revolutionary and Transformative frames), his words may be respected by many readers who are aware of Apple's success and/or are owners of Apple products. Therefore, the use of his opinion may give such evaluations (i.e. VR being isolating) significant weight to readers (Go, Jung and Wu, 2014). Thus, whereas Cook acts as a frame advocate for AR being social, he acts as a frame critic for VR being social.

Nevertheless, while it is useful to note that some of the news discourse counters the Social frame, it must be reiterated that the news articles were much more likely to focus on the social aspects of XR rather than to present it as isolating. Since framing deals with salience, or "making a piece of infor-mation more noticeable, meaningful, or memorable to audiences" (Entman, 1993: 53), it is clear that the Social frame was much more powerful in the news coverage of XR than its counterpart (isolating). Previous studies found that moral panics were created around videogames regarding violence and social isolation (Rogers, 2013). Despite videogames being the most mentioned application within the news articles, this moral panic does not extend to XR. This is surprising since the news articles focused on VR, which, as opposed to AR or MR, covers the user's view of the real world, literally isolating them from their surroundings. The news coverage of XR goes against previous moral panic trends, which is similar to De Keere, Thunnissen and Kuipers' (2020) findings regarding binge-watching discussed in Chapter 2. Instead, XR news includes repeated positive frames that support XR companies' marketing efforts.

Indeed, the preference for the positive Social frame over representing XR as isolating improves the perception of the compatibility of XR. Rogers states that compatibility is "the degree to which an innovation is perceived as consistent with the existing values, past experiences, and needs of potential adopters" (2003: 240). A technology that is social is much more compatible with the existing values and needs of potential adopters than one that causes isolation. Since higher compatibility leads to higher rates of adoption (Rogers, 2003), using the Social frame as opposed to an isolating one could support the diffusion of XR. Furthermore, the fact that this Social frame appears in the news and marketing discourses increases its strength as each reinforces the other. The news acts a promotional tool for XR companies and benefits their commercial interests.

Easy to Use

While the Social frame focused on the software aspect of XR, the rest of this chapter discusses those frames that are related to XR hardware. Despite XR being framed as Advanced and High-Quality (see Chapter 5), the technology was also framed as Easy to Use. The current section examines the framing devices used to construct the Easy to Use frame, making comparisons between the news articles and marketing materials.

To start, 12 percent of news articles included words in the "easy to use" category. Although this is not the majority of articles, in comparison, terms in the "difficult to use" group appeared in just 2 percent of articles overall – a much lower portion. Furthermore, every news outlet used words referring to XR as "easy to use" more than those in the "difficult to use" category. Similarly, whether articles focused on VR or AR/MR, terms in the "easy to use" group were always used more than those in the "difficult to use" category (see Appendix 7). These figures show that the news coverage has consistently favoured a positive frame (Easy to Use) over a negative one (e.g. complex or difficult to use), regardless of news outlet or XR type. When frames are repeated in different sources, their persuasive power increases (Nordfors, 2009). This means that the Easy to Use frame could have particular influence on how readers view XR.

However, it should be noted that there were some important differences between the news outlets regarding the Easy to Use frame. While terms in the "easy to use" category appeared in the *Guardian* and *MailOnline* in a similar portion of articles (11 and 12 percent respectively) only two articles in the *Sun* (3 percent) included these words. Furthermore, only one article in the *Sun* used a term from the "difficult to use" category. This shows that the topic of ease of use (whether easy or difficult) was very rarely mentioned in this outlet. Indeed, since frames are usually persistent (Gitlin, 1980), such few uses of words from the "easy to use" category in the *Sun* suggests that the Easy to Use frame was absent from this news outlet. This is one of the few instances in which a frame did not appear in all three news outlets to some extent. In this way, the media organisation factor (Shoemaker and Reese, 2014) appears to have affected the frame-building of XR quite strongly regarding this particular frame.

To a lesser extent, examining the use of words in the "easy to use" category across the sample period shows that this frame somewhat varied in emphasis over the years. The number of articles using terms from the "easy to use" category fluctuated across time, with two peaks in 2013 (24 percent) and 2015 (19 percent). Since 2016 was the major year for product releases (Steinicke, 2016), audiences would have been in the knowledge-building stage of Rogers' (2003) innovation-decision process in 2013 and 2015. Therefore, framing XR as Easy to Use at this stage could have created an initially positive attitude towards the technology. However, the reduction in the appearance of "easy to use" words in 2016 and 2017 suggests that this frame became less common once several XR products had been released into the market. That is to say, the appearance of the Easy to Use frame differed depending on the development stage of XR. Nevertheless, this reduction of "easy to use" words in the later years of the sample does not mean that XR was then framed as difficult to use. Words in the "easy to use" category appeared more than those from the "difficult to use" group in every year. Thus, although the strength of the Easy to Use frame varied, it was never overtaken by a counterframe. Combined with the other frames previously discussed, this highlights a lack of critical representations of XR. Since the news is the public's main source of information about emerging technologies (Whitton and Maclure, 2015; Williams, 2003), this could lead readers to view it in a positive light.

As well as specific word choices, modifiers were used as rhetorical framing devices to increase the strength the Easy to Use frame. In particular, the apparently natural feel of XR interaction was highlighted. The stem *natural** was the second most used term within the "easy to use" category. Describing interaction as natural implies there is no effort needed to understand how it works – it is easy and intuitive, thus exhibiting the Easy to Use frame. Examining this word in context within the news articles demonstrates that the idea of being able to naturally interact with the XR environment was emphasised. For instance, a journalist in the *Guardian* writing about his experience of a PlayStation VR demo claimed: "It all feels very natural and intuitive, holding the trigger button to grip the swords and then swinging them the way you would a real sword" (Gibbs, 2014). Firstly, he enhances the apparently natural feel of the experience with the modifier "very". The journalist then explains the controller can be used in the same way a real sword would be, again suggesting it offers a natural way to interact with the virtual environment.

Emphasising this idea even further, the following quotation appeared in three *MailOnline* articles: "Rory Abovitz, Magic Leap's CEO, said his firm is working on 'the most natural and human-friendly wearable computing interface in the world'" (Griffiths, 2014; Griffiths and Prigg, 2014, 2015). Instead of simply arguing the experience is "very" natural, the superlative "most" strengthens the Easy to Use frame even more. Additionally, since this quote was repeated in three articles, the salience of the message depicting the Easy to Use frame is increased. Therefore, as has been found with other frames, a frame advocate (relating to the social institutions factor (Shoemaker and Reese, 2014)) has again been given substantial power to portray XR in a way that is desirable to them, due to the routine practice of the *MailOnline* to repeat sections of its articles. This is likely caused by the wider social system leading news organisations to produce content quickly for commercial gain.

Regarding the relationship between XR news and marketing, it was found that the idea of natural interaction was also used in the promotional materials to create the Easy to Use frame. Firstly, since the above quotation came from the CEO of Magic Leap, it is not surprising that this idea was present in Magic Leap marketing. This could be seen on their website which explained that the interface of the device "provides the tools needed to break free from outdated conventions of point and click interfaces, delivering a more natural and intuitive way to interact with technology" (Magic Leap, 2017). Similarly, a repeated phrase within HoloLens marketing was that it creates a "more natural way to interact" (e.g. Microsoft, 2015a), in comparison to traditional ways of interacting with technology. In VR marketing, this idea of natural interaction was also present when referencing the controllers used. The Gear VR website noted that "[c]ontrol comes naturally" (Samsung, 2017). Additionally, the Oculus Rift website stated: "Before you even pick up a pair of Touch controllers, you know how to use them. Intuitive actions in VR feel as natural as using your real hands" (Oculus VR, 2016). In this example, the Easy to Use frame is highlighted by claiming no practise or effort is needed, since an individual will feel as if they are simply interacting with the virtual environment using their own hands. These excerpts from the marketing demonstrate that the Easy to Use frame is shared between the two discourses to the extent that the same framing device has been used to construct it, thus reinforcing the apparent facticity of the frame (Van Gorp, 2007).

Other framing devices were also uncovered. In a similar way to frames already discussed, a quotation from Mark Zuckerberg was used as a technical framing device to portray XR as Easy to Use. The following quote first appeared in the *Guardian* and the *MailOnline* on the same date (25 March 2014):

Imagine enjoying a court side seat at a game, studying in a classroom of students and teachers all over the world or consulting with a doctor face-to-face – just by putting on goggles in your home.

(Kiss, 2014; Prigg, 2014a)

The use of the modifier "just" in the last sentence of this quotation implies that each of these virtual experiences will be quickly and easily accessible. This quotation appeared in 11 different articles in the *Guardian* and eight *MailOnline* articles. As in other instances of repeated quotations, this finding suggests that a frame advocate (a social institutions level influence (Shoemaker and Reese, 2014)) from an XR company has again been given the power to define XR in positive terms. Frame advocates "select and enhance" certain aspects of a topic "to promote their own interests" (Hallberg-Sramek, Bjärstig and Nordin, 2020: 200). Using frame advocates as sources legitimises their chosen frame (Geiß, Weber and Quiring, 2016), thus giving these sources significant power in defining an issue. Furthermore, while it has already been established that the *MailOnline* copies and pastes sections of its previous articles into new ones (thus sometimes resulting in the repetition of certain quotes), the *Guardian* does not do this. Therefore, the repetition of this statement is significant because it shows that *Guardian* journalists have chosen to include this quotation in at least 11 different articles. With this in mind, it appears the *Guardian* in particular has put considerable emphasis on the Easy to Use frame by reinforcing the views of an XR company owner.

In a different way, another rhetorical framing device used to construct the Easy to Use frame was emphasising the immediacy of certain processes. This was applied to XR hardware as well as software. Regarding the software, one of the earliest articles in the *MailOnline* about Google Glass noted that users "could be given information instantly on the buildings they are looking at, on nearby landmarks or friends who are in the area" (Narain, 2012). The idea of instant information implies that using the device is easy and convenient. Similarly, a *MailOnline* article about an AR application that can translate foreign signs explained: "When a user looks at foreign writing, it is translated in real-time" (Woollaston, 2013). The use of "real-time" suggests this translation happens immediately. Furthermore, this is said to happen by simply *looking* at foreign text, which appears much more convenient than typing the words into a translator or referencing a word in a dictionary. In a study of AR in e-commerce settings, Kannaiah and Shanthi (2015) found that the desire for instant information was one of the factors that attracted consumers to AR. Therefore, framing XR as Easy to Use in this way could also make the technology seem more appealing to readers.

Regarding the hardware, this framing device was used in relation to smartphone-based VR headsets when describing how the products are set up. For instance, a *MailOnline* article about Google Daydream noted: "you open a hatch, pop the phone in, and suddenly you're fishing or exploring the world in VR via Street View" (Waugh, 2017). The word "pop" has connotations that imply the process can be done with little effort. This is emphasised by stating that, as soon as the phone is in the headset, the user is "suddenly" in the virtual world, insinuating the process is not only easy but can be done very quickly. Similarly, another article about Gear VR stated that "users slip a Note 3 tablet into the $200 headset to provide the screen" (Prigg, 2014b). The use of "slip" has similar connotations to "pop", suggesting the process is very easy.

Importantly, Gear VR promotional content also used this framing device. The Gear VR website stated: "Just snap your phone into the Gear VR and you're in virtual reality" (Oculus VR, 2015). Using the words "just" and "simply" implies it is easy and simple to do this. Additionally, a Facebook post about Gear VR from Samsung Global read: "Click in, boot up and start exploring" (Samsung Global, 2016). Using four one syllable words (click, in, boot, up) to explain the setup process reads quickly, creating the impression that the setup itself is quick and easy. This is further evidence to suggest the Easy to Use frame is shared between the news and marketing of XR.

As well as framing XR as Easy to Use by highlighting immediacy, the hands-free nature of some devices was referenced to construct this frame. For example, a journalist writing about HoloLens gestures in the *Guardian* stated: "I just have to lift my hand up in front of the device's sensors, raise a finger then make a sort of clicking gesture, like pressing the button on a mouse" (Stuart, 2015). The use of "just", combined with relating this process to an action many readers will be familiar with (clicking a mouse) enhances the idea that it is easy to do. Similarly, a *MailOnline* article about the Fove headset stated in its opening line: "No more fiddling with remote-controller buttons or a mouse. Just look" (Associated Press, 2016). The traditional way of interacting with technology (buttons/mouse) is classed as "fiddly", whereas being able to use the eyes to interact is said to be as simple as "just looking". The idea of using the eyes to interact with ease was also highlighted in a *MailOnline* article about the RideOn VR headset (which uses eye tracking technology):

> Worn while skiing or snowboarding, RideOn wearers will be able to message friends **in the blink of an eye**, stream live skiing videos and ride through virtual slalom tracks chasing their favourite ski athletes down the mountain, **without pressing a single button.** [...] **No external devices, phone apps, or voice activation is necessary.** Instead, wearers look at icons fixed to the sky, their friends, or points of interest.
>
> (Sturgis, 2015)

Specifically, the sections highlighted in bold contribute to framing the device as Easy to Use. Actions can be carried out quickly and easily ("in the blink of an eye") without the need for pressing any buttons or using additional software/hardware. In particular, this framing device highlights the relative advantage of XR by claiming it is better than already established technologies. The greater the perceived relative advantage, the more likely consumers are to adopt an innovation (Rogers, 2003). Thus, highlighting the Easy to Use frame in this way could support XR's adoption.

Regarding the connection with XR promotional content, Google Glass marketing in particular noted the hands-free features of the device to frame it as Easy to Use. It was emphasised that Google Glass can be used hands-free, such as for accessing information while carrying out fitness activities and viewing recipes while cooking (Google, 2013b). Similarly, in a video demonstrating one user's experience of Google Glass (Google, 2014), the user receives a call from his mother while cooking. He is able to answer the call without using his hands so that he can continue cooking while speaking. This highlights the convenience of being able to interact with the device hands-free. Furthermore, the same idea is emphasised in a tweet from the Google Glass Twitter account: "Catch all your phone notifications without having to pull that Android out of your pocket" (Google Glass, 2014). Again, the hands-free nature of the device is shown to make it easy to use, this time in comparison to a standard smartphone. The idea persisted into the enterprise rebrand of Google Glass when the new website stated: "Glass is a hands-free device, for hands-on workers" (Google, 2017). The website continued by noting several examples of how the device can be used by workers without the need to interact with it using their hands. Therefore, in these instances, just as in the news coverage, the way interaction with XR products has been described has contributed to framing it as Easy to Use.

As has been considered with the other frames discussed throughout this book, it is important to acknowledge any attempts at countering the Easy to Use frame. This frame was contested by arguing against the idea that interaction with XR products is natural (as discussed above). For example, one *MailOnline* article from 2013 stated that "lunging forward with your head to move forward" (*MailOnline*, 2013) is the way a user must interact with Oculus Rift. Having to "lunge forward" appears to be a very unnatural way of interacting with the virtual environment, portraying it as an experience that is difficult to control. Similarly, before the Gear VR headset had its own controller, interaction was also explained negatively in this *MailOnline* article: "With the Gear VR, you have to move your head to point a cursor at something, then reach for a button on the headset – which is far from ideal" (Prigg, 2016). Still, although it is significant that there were some instances of opposing the Easy to Use frame, it should be remembered that the stem *natural** was used 49 times within the news articles, whereas *unnatural** only appeared twice. Therefore, articles were much more likely to highlight ease of use than difficulty of use when reporting on XR. This is another example of the news discourse favouring positive frames over those that are negative.

The appearance of the Easy to Use frame in XR news and marketing shows some similarities with research of other technologies. For instance, Kang, Lee and De La Cerda (2015) found that ease of use was one of the two most common frames used in US television news of the smartphone. Additionally, in their study of news about Twitter, Arceneaux and Weiss (2010) found that articles often emphasised the near-instant dissemination of information using this platform, which relates to one of the framing devices that contributed to constructing the Easy to Use frame in XR news. Similarly, Therrien and Lefebvre uncovered an accessibility frame in their study of videogame marketing which was manifested "through an emphasis on ergonomic controls, a lenient learning curve or the presence of adjustable/easy difficulty levels" (2017: 56). In this case, it seems that XR news coverage has more in common with videogame marketing than videogame news which has been found to include moral panic style content (Rogers, 2013). Furthermore, the fact that ease of use is a topic within XR news at all indicates that journalists are targeting this content towards potential consumers of the technology as usability is something that would only be of interest to those considering purchasing a device. This demonstrates another similarity between XR news and lifestyle journalism which treats the audience as consumers (Hanusch, 2012).

The Easy to Use frame not only presents XR positively, but also relates to the complexity attribute of an innovation. Rogers defines this attribute as "the degree to which an innovation is perceived as difficult to understand and use" (2003: 16). Furthermore, innovations "that are simpler to understand are adopted more rapidly than innovations that require the adopter to develop new skills and understandings" (2003: 16). Based on Rogers' assessment, framing XR as Easy to Use could lead it to be adopted more quickly. In other words, the news articles have again promoted the diffusion of XR by using this frame. Just like the other frames already discussed, the appearance of the Easy to Use frame in the news sample provides further evidence to suggest the news supports the commercial interests of XR companies.

Comfortable

The final section in this chapter discusses another way XR was framed relating to the user experience of the hardware itself: Comfortable. This refers to physical comfort surrounding the design of the hardware, rather than psychological comfort. In what follows, the framing devices used to create this frame in XR news are closely examined, alongside a comparison with the marketing materials.

Considering word usage to begin with, terms in the "comfortable" category appeared 208 times in 11 percent of articles overall. As was the case with the Easy to Use frame, words that could counter the Comfortable frame rarely appeared. Terms in the "uncomfortable" category were used just 33 times in 3 percent of articles. This means that words relating to XR being comfortable were used over six times more than those referring to it as uncomfortable. Moreover, both VR and AR/MR articles used "comfortable" words considerably more than "uncomfortable" words, with no notable differences between the portion of articles they appeared in. Similarly, despite the use of words in the "comfortable" category fluctuating over the years of the sample, they were always used much more than those in the "uncomfortable" category. Therefore, the Comfortable frame was consistent regardless of year or type of XR.

However, it should be noted that the Comfortable frame does not appear to be prevalent in all news outlets. In fact, the *MailOnline* was the only news outlet to use words from the "comfortable" category in a substantial portion of articles (14 percent) and much more than words in the "uncomfortable" group (2 percent). On the other hand, the *Sun* very rarely mentioned words in either of these categories (two uses from each), meaning discussions of comfort, or indeed discomfort, were not common in the *Sun* articles. Similarly, the percentage of articles in the *Guardian* using words in the "comfortable" and "uncomfortable" categories were near identical (6 and 5 percent respectively). Considering these figures, it seems that the *MailOnline* has been the main proponent of the Comfortable frame in comparison to the other news outlets. Furthermore, just as the Easy to Use frame was absent from the *Sun*, it appears this publication did not use the Comfortable frame either. In this way, the media organisation factor (Shoemaker and Reese, 2014) has again impacted whether a frame was present or not. While this perhaps means the Comfortable frame reaches a smaller audience, it is still significant that it appeared in the *MailOnline*. To specify, the *MailOnline* was the world's most read online news source during the sample period of this study (Greenslade, 2012; Johnston, 2018; *This is Money*, 2016), meaning its individual readership was very large. For this frame to appear in the *MailOnline* means that it could still encourage a large group to view XR favourably.

Turning to the use of specific words, in the "comfortable" category, the term *light(weight)* was the most common and this lightness was emphasised in some articles. For example, a *MailOnline* article claimed: "The Playstation VR headset has been designed to be as light and as comfortable as possible" (Best, 2016), suggesting the comfort is very high. More extremely, the lightness of Gear VR was exaggerated in another *MailOnline* article: "At first glance the headset looks like it will be really heavy, but it just feels like a pair of goggles you used to wear when doing science at school" (Shoffman, 2014). While even the lightest version of the Gear VR headset is 345 grams (Samsung, n.d.), the journalist compares the headset to protective plastic goggles worn during science classes. Gear VR is significantly heavier than a pair of science glasses, making this an exaggeration. However, since the majority of generalist news readers "do not have technical knowledge or background" (Weiss-Blatt, 2016: 415), they may not be aware that this is exaggerated. This, then, creates the impression that Gear VR is extremely lightweight, thus framing it as Comfortable.

Similarly, other articles constructed the Comfortable frame by focusing on headset weight distribution. Regarding PlayStation VR, the *Guardian* stated:

The new design places the device's weight on the top of the head so that there's little pressure on top of face [*sic*] – a mild dig at Oculus Rift's more intrusive goggle-style design. The headset is comparatively light, and the redesigned strap distributes the weight evenly for a comfortable fit.

(Parkin, 2015)

In this example, PlayStation VR is said to be more comfortable than Oculus Rift, due to its distribution of weight on the top of the head. While PlayStation VR marketing was not analysed, it was found that the distribution of weight to improve comfort was highlighted on the HoloLens website:

Designed for comfort.
    The headband is designed like a performance car with great weight distribution for a comfortable fit. Weight is distributed around the crown of your head, saving your ears or nose from undue pressure.

(Microsoft, 2015b)

Here, the comfort of this device is portrayed to be extremely high-quality by comparing it to a "performance car". The explanation of how the headset distributes weight further enhances this. Additionally, the marketing of various XR devices described them as "light" or "lightweight" (e.g. Google, 2013a; Magic Leap, 2017; Microsoft, 2015a; Oculus VR, 2012; Samsung, 2017). Therefore, it is clear that the Comfortable frame is shared between the two discourses.

However, there were some instances where the Comfortable frame was opposed within the news articles. Firstly, in a report from the *Guardian*, a journalist highlighted their discomfort while using the Google Cardboard headset: "I am intensely aware that the bridge of my nose is being assaulted by the hard edges of the headset" (Shubber, 2014). The use of the modifier "intensely" puts great emphasis on their discomfort. Moreover, describing the edges of the headset as "hard" and using the metaphor "assaulted" implies a very uncomfortable experience. Since the Cardboard headset is literally made of cardboard, it is not surprising that this product in particular has been singled out as being uncomfortable. Still, another journalist from the *Guardian* did highlight his discomfort when using an early version of Oculus Rift: "Wearing it felt like having uncomfortable ski goggles clamped to my face" (Hern, 2015). Aside from the use of "uncomfortable", "clamped" suggests a very tight fit. Though this presents the device as highly uncomfortable, the statement was contrasted in the next paragraph with the journalist's recent experience of Oculus Rift. He claimed "I finally saw what the fuss was about" (Hern, 2015), with no further mention of discomfort. Therefore, the discomfort of XR was not emphasised to the extent that it would be classified as a frame. This is supported by the numerical data discussed above which shows that the Comfortable frame was favoured by the news outlets.

The preference for the Comfortable frame demonstrates another example of the news coverage focusing on positive rather than negative frames when it comes to XR. As previously noted, the news is the main source of information about emerging technologies for the general public (Whitton and Maclure, 2015; Williams, 2003). Thus, combined with the other positive frames already discussed, this is further evidence to suggest that the news fosters a positive attitude towards XR, which could lead readers to be more likely to purchase these products. While the Comfortable frame does not directly relate to any of Rogers' (2003) perceived attributes of innovations, it does coincide with one of Buenaflor and Kim's (2013) factors concerning the acceptability of wearable computers. According to the authors, "[p]hysical comfort and safety is an essential consideration" when it comes to the acceptance of wearable computers (2013: 109). They also insist that "a user's perception of a new technology significantly affects acceptance" (2013: 107). Therefore, as a wearable technology, the emphasis of the Comfortable frame in the news coverage could contribute to supporting the adoption of XR. Presenting XR in such a way is clearly of benefit to XR companies since this frame also appears in their marketing. Thus, the news acts as a promotional tool for XR by using this frame, supporting the capitalist ideologies of XR companies. Like the Easy to Use frame, the Comfortable frame treats the audience as consumers, as is the case in lifestyle journalism (Hanusch, 2012), since the comfort of the device would not be of interest to someone unless they were considering purchasing it. Therefore, although these articles are presented as news, they include characteristics of lifestyle journalism which treats audiences as consumers.

Marketing Influence on News Discourse

Considering these results alongside those presented in Chapters 4 and 5 shows that eight frames that were present in the marketing of XR also appeared in the news coverage. However, it should also be noted that some frames were identified in the marketing that did not appear in the news coverage. Firstly, a Personal frame appeared in the marketing of some products to present the technology as allowing a tailored, individual experience. Secondly, a Boundless frame in the marketing implied there is a plethora of content available for XR. Finally, a Magical frame was found in the marketing of (unsurprisingly) Magic Leap, with its tagline being to "bring magic back into the world". This shows that there were some differences between the two discourses and the news did not use *all* of the frames present in XR marketing. Nevertheless, the major finding here is that several more frames *were* shared between the two samples (eight), than those that were unique to the marketing (three).

This indicates that, within the social institutions level of the hierarchy of influences (Shoemaker and Reese, 2014), the marketing materials of XR have been instrumental in the framing process for XR news. Without analysing journalists in the newsroom, it cannot be certain whether they have been directly influenced by this marketing when creating XR news, but the strong similarities in the way XR is framed in the two discourses certainly supports this claim. Furthermore, the prevalence of XR company owners and content creators as sources in the news articles, as well as the repetition of quotes from company owners such as Zuckerberg as framing devices, shows that these groups have been given significant power to frame XR. Whether journalists have been influenced by these groups or not, when identical frames appear in different media, the persuasive power of the frames is enhanced "because the media appear to address the audience with a single voice" (Van Gorp, 2007). That is to say, the discourses work to reinforce each other. Since the way XR is framed in its marketing is designed to sell the products, the fact that these same frames also exist in the news content suggests that the news is also acting as a promotional tool to support the diffusion of the technology. Although the journalistic norm of objectivity states that promotional content should be clearly separate from news content (Carlson, 2017), this does not seem to have happened in relation to XR. Instead, this hints at the marketisation (Fairclough, 1993) of XR news and the blurring of news and promotional content, supporting Chyi and Lee's (2018) argument that technology news is commercialised.

Moreover, each of the frames in this chapter emphasised a positive aspect of the XR experience, rather than taking a critical stance, which is in line with the findings discussed in previous chapters. As each frame emphasises an aspect of an innovation or technology that can increase the chance of adoption, the use of these frames promotes the diffusion of XR. This is concerning regarding the state of technology news (at least about XR) as it shows journalists are not performing their fourth estate role. Instead of providing the public with information about the potential benefits and risks of XR, they only focus on positively emphasising aspects of the technology that increase the likelihood of adoption. Added to this, these are the same frames that appear in XR marketing, meaning XR companies perceive these traits to be the key factors that will help them sell their products. As these frames are repeated in the news articles, the discourse that is supposed to critically inform the public (Fjæstad, 2007) instead promotes products and persuades audiences to purchase these devices. In other words, the news prioritises the interests of XR companies rather than the general public. The next chapter provides further insight into this by examining the evaluative frames applied to XR in the news specifically, as well as the overall tone of the coverage.

References

Arceneaux, N. and Weiss, A.S. (2010) 'Seems Stupid Until You Try It: Press Coverage of Twitter, 2006–9', *New Media & Society*, 12(8), pp. 1262–1279. doi:10.1177/1461444809360773.

Associated Press (2016) 'The end of the joystick? VR headset with built in eye tracking 'adds empathy' to virtual worlds', *MailOnline*, 17 March. Available at: www.dailymail.co.uk/sciencetech/article-3496385/Startup-makes-virtual-reality-intuitive-eye-tracking.html (Accessed: 12 February 2019).

Best, S. (2016) 'PlayStation virtual reality is here: Sony announces new £350 VR headset for PS4 will be available from October', *MailOnline*, 14 June. Available at: www.dailymail.co.uk/sciencetech/article-3640169/PlayStation-VR-headset-hit-market-October.html (Accessed: 12 February 2019).

Best, S. (2017) 'Apple's Tim Cook predicts augmented reality will be bigger than VR because it doesn't isolate people in their own worlds', *MailOnline*, 12 October. Available at: www.dailymail.co.uk/sciencetech/article-4973026/Tim-Cook-SLAMS-virtual-reality-new-interview.html (Accessed: 7 February 2019).

Brigham, T.J. (2017) 'Reality Check: Basics of Augmented, Virtual, and Mixed Reality', *Medical Reference Services Quarterly*, 36(2), pp. 171–178. doi:10.1080/02763869.2017.1293987.

Buenaflor, C. and Kim, H. (2013) 'Six Human Factors to Acceptability of Wearable Computers', *International Journal of Multimedia and Ubiquitous Engineering*, 8(3), pp. 103–114.

Carlson, M. (2017) *Journalistic Authority: Legitimating News in the Digital Era*. New York: Columbia University Press.

Chyi, H.I. and Lee, A.M. (2018) 'Commercialization of Technology News', *Journalism Practice*, 12(5), pp. 585–604. doi:10.1080/17512786.2017.1333447.

De Keere, K., Thunnissen, E. and Kuipers, G. (2020) 'Defusing Moral Panic: Legitimizing Binge-Watching as Manageable, High-Quality, Middle-Class Hedonism', *Media, Culture & Society*, [Preprint], pp. 1–19. doi:10.1177/0163443720972315.

Dredge, S. (2016a) 'The complete guide to virtual reality — everything you need to get started', *The Guardian*, 10 November. Available at: www.theguardian.com/technology/2016/nov/10/virtual-reality-guide-headsets-apps-games-vr (Accessed: 20 December 2018).

Dredge, S. (2016b) 'Three really real questions about the future of virtual reality', *The Guardian*, 7 January. Available at: www.theguardian.com/technology/2016/jan/07/virtual-reality-future-oculus-rift-vr (Accessed: 20 December 2018).

Entman, R.M. (1993) 'Framing: Towards Clarification of a Fractured Paradigm', *Journal of Communication*, 43(4), pp. 51–58.

Evans, C.L. (2016) 'Virtual reality may look cool, but it will feel empty without community', *The Guardian*, 23 August. Available at: www.theguardian.com/commentisfree/2016/aug/23/virtual-reality-cool-empty-community-oculus-rift (Accessed: 21 December 2018).

Facebook (2014) *Facebook to Acquire Oculus*. 25 March [Press Release]. Available at: https://newsroom.fb.com/news/2014/03/facebook-to-acquire-oculus/ (Accessed: 31 October 2019).

Fairclough, N. (1993) 'Critical Discourse Analysis and the Marketization of Public Discourse: The Universities', *Discourse & Society*, 4(2), pp. 133–168.

Fjæstad, B. (2007) 'Why Journalists Report Science as They Do', in Bauer, M.W. and Bucci, M. (eds.) *Journalism, Science and Society: Science Communication between News and Public Relations*. Oxon: Routledge, pp. 123–131.

Geiß, S., Weber, M. and Quiring, O. (2016) 'Frame Competition after Key Events: A Longitudinal Study of Media Framing of Economic Policy after the Lehman Brothers Bankruptcy 2008–2009', *International Journal of Public Opinion Research*, 29(3), pp. 471–496. doi:10.1093/ijpor/edw001.

Gibbs, S. (2014) 'Sony's Project Morpheus brings virtual reality to mainstream console gaming', *The Guardian*, 12 May. Available at: www.theguardian.com/technology/2014/may/12/sonys-project-morpheus-virtual-reality-console-gaming (Accessed: 19 December 2018).

Gitlin, T. (1980) *The Whole World Is Watching*. Berkeley: University of California Press.

Go, E., Jung, E.H. and Wu, M. (2014) 'The Effects of Source Cues on Online News Perception', *Computers in Human Behavior*, 38, pp. 358–367. doi:10.1016/j.chb.2014.05.044.

Google (2013a) *What It Does*. Available at: https://web.archive.org/web/20130223025 434/http:/www.google.com/glass/start/what-it-does/ (Accessed: 3 September 2019).

Google (2013b) *Glass*. Available at: https://web.archive.org/web/20131120024937/http:/www.google.com/glass/start/ (Accessed: 3 September 2019).

Google (2014) *How It Feels*. Available at: https://web.archive.org/web/20140320214 630/http:/www.google.com/glass/start/how-it-feels/ (Accessed: 3 September 2019).

Google (2017) *Glass*. Available at: https://web.archive.org/web/20170718142617/http:/www.x.company/glass/ (Accessed: 3 September 2019).

Google Glass (2014) 'Catch all your phone notifications without having to pull that Android out of your pocket' [Twitter] 12 October. Available at: https://twitter.com/googleglass/status/522156397899702272 (Accessed: 13 September 2019).

Greenslade, R. (2012) 'Mail Online Goes Top of the World', *The Guardian*, 25 January. Available at: www.theguardian.com/media/greenslade/2012/jan/25/dailymail-internet (Accessed: 8 October 2020).

Griffiths, S. (2014) 'Google to take on Facebook's Oculus Rift: firm set to spend $500m on 'cinematic reality' firm Magic Leap', *MailOnline*, 14 October. Available at: www.dailymail.co.uk/sciencetech/article-2792411/google-set-spend-500m-cinematic-reality-firm-magic-leap-head-head-facebook-s-oculus-rift.html (Accessed: 14 February 2019).

Griffiths, S. and Prigg, M. (2014) 'What IS Magic Leap? Secretive firm raises $542m from Google and others for its 'cinematic reality' system – but still won't reveal what the technology is', *MailOnline*, 21 October. Available at: www.dailymail.co.uk/sciencetech/article-2802333/what-magic-leap-secretive-firm-raises-542m-google-cinematic-reality-won-t-reveal-technology-is.html (Accessed: 14 February 2019).

Griffiths, S. and Prigg, M. (2015) 'Magic Leap set to revolutionise every aspect of daily life: patent of 'secret' augmented reality headset reveals uses in shops, hospitals and homes', *MailOnline*, 19 January. Available at: www.dailymail.co.uk/sciencet ech/article-2916696/Magic-Leap-set-revolutionise-aspect-daily-life-Patent-secret-augmented-reality-headset-reveals-uses-shops-hospitals-homes.html (Accessed: 14 February 2019).

Hallberg-Sramek, I., Bjärstig, T. and Nordin, A. (2020) 'Framing Woodland Key Habitats in the Swedish Media — How Has the Framing Changed Over Time?', *Scandinavian Journal of Forest Research*, 35(3–4), pp. 198–209. doi:10.1080/02827581.2020.1761444.

Hanusch, F. (2012) 'Broadening the Focus: The Case for Lifestyle Journalism as a Field of Scholarly Inquiry', *Journalism Practice*, 6(1), pp. 2–11. doi:10.1080/17512786.2011.622895.

Hern, A. (2015) 'Will 2016 be the year virtual reality gaming takes off?', *The Guardian*, 28 December. Available at: www.theguardian.com/technology/2015/dec/28/virtual-reality-gaming-takes-off-2016 (Accessed: 21 December 2018).

Johnston, C. (2018) 'Paul Dacre Stepping Down as Daily Mail Editor', *BBC*, 6 June. Available at: www.bbc.co.uk/news/business-44391449 (Accessed: 8 October 2020).

Kang, S., Lee, K.M. and De La Cerda, Y. (2015) 'U.S. Television News about the Smartphone: A Framing Analysis of Issues, Sources and Aspects', *Online Journal of Communication and Media Technologies*, 5(1), pp. 174–196.

Kannaiah, D. and Shanthi, R. (2015) 'The Impact of Augmented Reality on E-Commerce', *Journal of Marketing and Consumer Research*, 8, pp. 64–73.

Kiss, J. (2014) 'Oculus: Facebook buys virtual reality gaming firm for $2bn', *The Guardian*, 25 March. Available at: www.theguardian.com/technology/2014/mar/25/facebook-buys-virtual-reality-gaming-firm-oculus (Accessed: 19 December 2018).

Liberatore, S. (2016) 'The HoloLens is here (if you have $3,000 to spare): Microsoft begins taking preorders from developers for augmented reality headset', *MailOnline*, 29 February. Available at: www.dailymail.co.uk/sciencetech/article-3469683/The-HoloLens-3-000-spare-Microsoft-begins-taking-preorders-developers-augmented-reality-headset.html (Accessed: 7 December 2018).

Magic Leap (2017) *Magic Leap*. Available at: https://web.archive.org/web/2017123 1203634/https:/www.magicleap.com/ (Accessed: 16 September 2019).

MailOnline (2013) 'The 3D goggles that'll have you believe you're living in The Matrix', 19 January. Available at: www.dailymail.co.uk/home/moslive/article-2263 921/Oculus-Rift-headset-3D-goggles-thatll-believe-youre-living-The-Matrix.html (Accessed: 20 February 2019).

Microsoft (2015a) *Microsoft HoloLens*. Available at: https://web.archive.org/web/ 20150124143645/http:/www.microsoft.com/microsoft-hololens/en-us (Accessed: 3 September 2019).

Microsoft (2015b) *Hardware*. Available at: https://web.archive.org/web/201 50728065657/http:/www.microsoft.com/microsoft-hololens/en-us/hardware (Accessed: 13 September 2019).

Narain, J. (2012) 'Surf the net, email, make calls – with your glasses! How the Google goggles work', *MailOnline*, 24 February. Available at: www.dailymail.co.uk/scie ncetech/article-2105628/Google-glasses-Surf-net-email-make-calls--Google-gogg les-work.html (Accessed: 20 February 2019).

Nordfors, D. (2009) 'Innovation Journalism, Attention Work and the Innovation Economy', *Innovation Journalism*, 6(1), pp. 1–46.

Oculus VR (2012) *About*. Available at: https://web.archive.org/web/20120609235 553/http:/oculusvr.com/?page_id=12 (Accessed: 30 August 2019).

Oculus VR (2015) *Gear VR*. Available at: https://web.archive.org/web/20151127171 834/https://www.oculus.com/en-us/gear-vr/ (Accessed: 2 September 2019).

Oculus VR (2016) *Rift*. Available at: http://web.archive.org/web/20161008154031/ https:/www3.oculus.com/en-us/rift/ (Accessed: 3 September 2019).

Pan, Z. and Kosicki, G.M. (1993) 'Framing Analysis: An Approach to News Discourse', *Political Communication*, 10, pp. 55–75.

Parkin, S. (2015) 'Sony Morpheus virtual reality headset to launch in first half of 2016', *The Guardian*, 4 March. Available at: www.theguardian.com/technology/ 2015/mar/04/sony-morpheus-playstation-4-ps4-virtual-reality-headset-to-launch-2016 (Accessed: 19 December 2018).

Prigg, M. (2014a) Facebook buys virtual reality headset firm Oculus for $2bn as Mark Zuckerberg promises to 'change the way we communicate', *MailOnline*, 25 March. Available at: www.dailymail.co.uk/sciencetech/article-2589367/Get-ready-social-platform-Facebook-buys-virtual-reality-firm-Oculus-2bn.html (Accessed: 14 February 2019).

Prigg, M. (2014b) 'Virtual reality is here: Samsung's $200 Gear VR headset goes on sale (although you'll need a $750 tablet to use it)', *MailOnline*, 8 December. Available at: www.dailymail.co.uk/sciencetech/article-2866088/Virtual-reality-Samsung-s-200-Gear-VR-headset-goes-sale-ll-need-750-tablet-use-it.html (Accessed: 14 February 2019).

Prigg, M. (2016) 'Google's Daydream: The $79 headset that could bring VR to the masses – if the apps can catch up', *MailOnline*, 10 November. Available at:

www.dailymail.co.uk/sciencetech/article-3924370/Google-s-Daydream-79-head set-bring-VR-masses-apps-catch-up.html (Accessed: 11 February 2019).

Rogers, E.M. (2003) *Diffusion of Innovations*. 5th edn. New York: Free Press.

Rogers, R. (2013) 'Critical Essay — Old Games, Same Concerns', *Technoculture*, 3. Available at: https://tcjournal.org/drupal/vol3/rogers (Accessed: 10 October 2016).

Samsung (2017) *Gear VR*. Available at: https://web.archive.org/web/20170330014 035/http:/www.samsung.com/global/galaxy-gear-vr/#!/ (Accessed: 29 August 2017).

Samsung (n.d.) *Gear VR Specifications*. Available at: www.samsung.com/global/gal axy/gear-vr/specs/ (Accessed: 11 March 2020).

Samsung Global (2016) 'Click in, boot up and start exploring everywhere from any-where' [Facebook] 14 January. Available at: www.facebook.com/SamsungGlobal (Accessed: 16 September 2019).

Samsung Mobile (2016) 'Celebrate together from a thousand miles away' [Twitter] 30 December. Available at: https://twitter.com/SamsungMobile/status/81481833755 3158144 (Accessed: 12 September 2019).

Shoemaker, P.J. and Reese, S.D. (2014) *Mediating the Message in the 21st Century*. Oxon: Routledge.

Shoffman, M. (2014) 'Wearable technology like Google Glass is tipped to become a multi-billion pound industry – but is it a good fit for investors?', *MailOnline*, 21 November. Available at: www.dailymail.co.uk/money/investing/article-2828 680/Wearable-technology-like-Google-Glass-tipped-multi-billion-pound-industry-good-fit-investors.html (Accessed: 14 February 2019).

Shubber, K. (2014) 'Are you ready for the virtual reality revolution?', *The Guardian*, 2 August. Available at: www.theguardian.com/technology/2014/aug/ 02/are-you-ready-for-virtual-reality-revolution-google-cardboard (Accessed: 13 December 2018).

Steinicke, F. (2016) *Being Really Virtual*. Cham, Switzerland: Springer.

Steuer, J. (1992) 'Defining Virtual Reality: Dimensions Determining Telepresence', *Journal of Communications*, 42(4), pp. 73–93.

Stuart, K. (2015) 'Minecraft on Hololens: the future of gaming is right in front of your eyes', *The Guardian*, 24 June. Available at: www.theguardian.com/technology/ 2015/jun/24/minecraft-hololens-mixed-augmented-reality-microsoft (Accessed: 19 December 2018).

Sturgis, I. (2015) 'Virtual reality SKIING! World's first true augmented reality ski goggles let adrenaline junkies create slalom tracks to follow and even video message friends on the slopes', *MailOnline*, 21 January. Available at: www.dailymail.co.uk/ travel/travel_news/article-2918199/Virtual-reality-SKIING-World-s-augmented-reality-ski-goggles-let-adrenaline-junkies-create-slalom-tracks-follow-video-mess age-friends-slopes.html (Accessed: 14 February 2019).

Therrien, C. and Lefebvre, I. (2017) 'Now You're Playing with Adverts: A Repertoire of Frames for the Historical Study of Game Culture through Marketing Discourse', *Kinephanos*, 7(November), pp. 37–73.

This is Money (2016) 'Mail online revenues up 27% as it maintains position as most popular English language newspaper website in the world', *This is Money*, 28 January. Available at: www.thisismoney.co.uk/money/markets/article-3421479/ Mail-Online-revenues-27-maintains-position-popular-English-language-newspa per-website-world.html (Accessed: 8 October 2020).

Van Gorp, B. (2007) 'The Constructionist Approach to Framing: Bringing Culture Back In', *Journal of Communication*, 57, pp. 60–78. doi:10.1111/j.1460-2466.2006.00329.x.

Waugh, R. (2017) 'Daydream (non)-believer: despite wider views and hundreds of new apps the Daydream 2 still fails to impress', *MailOnline*, 28 October. Available at: www.dailymail.co.uk/home/event/article-5021123/The-Daydream-2-fails-impress.html (Accessed: 7 February 2019).

Weiss-Blatt, N. (2016) 'Tech Bloggers vs. Tech Journalists in Innovation Journalism', *Proceedings of the 3rd European Conference on Social Media Research*, Caen, France, 12–13 July, pp. 415–423.

Whitton, N. and Maclure, M. (2015) 'Video Game Discourses and Implications for Game-based Education', *Discourse: Studies in the Cultural Politics of Education*, pp. 1–13. doi:10.1080/01596306.2015.1123222.

Williams, D. (2003) 'The Video Game Lightning Rod', *Information, Communication & Society*, 6(4), pp. 523–550. doi:10.1080/1369118032000163240.

Woollaston, V. (2013) 'Is this the best holiday gadget ever? The Google Glass-style visor that translates ANY foreign menu and sign immediately', *MailOnline*, 30 September. Available at: www.dailymail.co.uk/sciencetech/article-2439232/NTT-Docomo-unveils-Google-Glass-style-visor-translates-foreign-menus-signs.html (Accessed: 6 December 2018).

Yadron, D. (2016) 'We've seen Magic Leap's device of the future, and it looks like Merlin's skull cap', *The Guardian*, 8 June. Available at: www.theguardian.com/technology/2016/jun/07/magic-leap-headset-design-patent-virtual-reality (Accessed: 21 December 2018).

7 Evaluative Framing of XR

The fourth and final frame category of the model for analysing media coverage of emerging technologies is Evaluation. Although some of the frames included in previous categories involve evaluating the technology in some sense (e.g. how comfortable the device is), this fourth frame category is broader. It allows a study to analyse the positive and negative portrayals of the technology in more depth without being restricted to the specifics of the other categories. As the perception of innovations can impact their success (Buenaflor and Kim, 2013), analysing how they have been evaluated in the news is key because this can shed light on the role the media plays in creating these perceptions and potentially impacting adoption. When analysing texts to identify frames in this category, there are a number of questions that researchers can ask:

- Overall, how positive/negative is the coverage?
- What is said to be good/bad about the technology?
- What benefits are mentioned and to what extent are they emphasised?
- What concerns are mentioned and to what extent are they emphasised?
- How important/significant is the technology deemed to be?

In relation to the study of XR news, four frames were identified in this category: Important; Successful; Affordable; and Much-Anticipated. This chapter discusses how each of these frames have been constructed before moving on to a wider analysis of positive and negative coverage of the technology.

Important

As discussed in Chapter 4, the Revolutionary and Transformative frame emphasised the importance of XR as able to create meaningful change. However, this importance was highlighted even more widely, resulting in the emergence of a specific Important frame within the news discourse. This involved presenting XR as a significant development with high importance.

DOI: 10.4324/9781003375814-7

The current section discusses the framing devices used to construct the Important frame in the news articles.

Starting by considering word usage, terms in the "important" category were used 179 times in 13 percent of articles. In comparison, words that could counter this frame, in the "trivial" category, were only used 51 times in 4 percent of articles. This shows that words presenting XR in a favourable light have been used more often than those that would do the opposite. However, inspecting the appearance of these words over time provides more nuanced insight into the use of this frame. At the start of the sample (2012), the percentage of news articles using words in both categories ("important" and "trivial") was the same: 8 percent. Thus, in the year that new XR products began being announced, the Important frame did not dominate the coverage. Instead, there were a mixture of viewpoints on this issue. Nevertheless, in every year after this, words in the "important" category were used in more articles than those in the "trivial" category. That is to say, from 2013 to 2017, the news outlets chose to use the Important frame to present XR favourably.

While there was not a significant difference between the use of these words in VR and AR/MR articles, every news outlet used words in the "important" category more than those in the "trivial" category, showing the dominance of the Important frame. The *Sun* did not use any words from the "trivial" category at all. Still, only 7 percent of its news items included words in the "important" group, meaning that this was not a very common frame in the *Sun*. In the *MailOnline*, 12 percent of articles included "important" words, whereas just 2 percent used terms from the "trivial" category, showing that the *MailOnline* placed particular emphasis on the Important frame over its potential counterpart. Alternatively, the *Guardian* used words in the "important" category the most (15 percent of articles) out of all news outlets. However, 11 percent of articles from this publication used terms from the "trivial" category. Thus, the *Guardian* appears to have taken a more balanced approach regarding the Important frame. Quality news outlets, such as the *Guardian*, are expected to adhere to the journalistic norms of objectivity and balance more than tabloids and middle-market publications (Bastos, 2019), which could explain this difference. Therefore, the variation between how often words were used in the three publications shows that the media organisation (factor three of the hierarchy of influences model (Shoemaker and Reese, 2014)) reporting on XR has had an impact on the strength of the Important frame.

Looking more closely at how often specific terms were used within the news articles highlights another noteworthy finding regarding the Important frame. Out of all words in the "important" category, *special* was the one to be used the most (7 percent of articles), even more than *importan** (e.g. important, importance) itself (2 percent). Bantimaroudis and Ban state that a "careful examination of word choices and the extent of their use in news coverage can reveal much about the organizing ideas, the framing choices, of the media" (2001: 177). Indeed, the frequent use of *special* does just that.

Whereas the use of the stem *importan** clearly denotes the Important frame, it is a less loaded term than a word such as *special*, which connotes importance while also implying that the way it is important is positive and perhaps different. Therefore, word choice has been used as a framing device not only to frame XR as Important, but to do this in a positive way. Because the news is the public's main source of information about emerging technologies (Whitton and Maclure, 2015; Williams, 2003) and the perception of an innovation is a key factor in its success (Buenaflor and Kim, 2013), this favourable evaluation of XR could promote its diffusion.

In addition to using specific words to frame XR as Important, other techniques also worked to present XR this way. Firstly, quotations were used as technical framing devices to construct this frame. In continuity with the findings discussed in previous chapters, quotes from both Mark Zuckerberg and Apple CEO Tim Cook were used to highlight this frame. In particular, Zuckerberg's statement that "we believe that VR is going to be the next big computing platform" (Reuters, 2016) or the "next major computing platform" (AFP, 2017) framed the technology as Important. Here, VR specifically is represented as highly significant as it is implied it will be a new way for people to interact with computers, rather than simply a gimmick. Variations of this quote were used in several news items: the phrase "next major computing platform" appeared in one *Guardian* article and five *MailOnline* articles. Similarly, the phrase "next big computing platform" appeared in five news articles, again with one from the *Guardian* and four in the *MailOnline*. Here, the *MailOnline*'s routine practice (Shoemaker and Reese, 2014) of copying and pasting parts of its articles has contributed to enhancing the Important frame. As noted previously, repetition of quotes could be a result of pressures on journalists to create news content quickly, thus indicating that this routine practice has developed due to commercial pressures in the social system, linking these two factors of the hierarchy of influences model.

Similarly, Tim Cook was also quoted to frame AR as Important. A *MailOnline* article stated that: "Apple CEO Tim Cook has called augmented reality (AR) a 'big idea' and people will 'have AR experiences every day, almost like eating three meals a day'" (Prigg, 2017). Comparing AR to the integral and everyday act of eating meals suggests it will be a big part of life, which highlights its significance. Furthermore, as mentioned in previous chapters, this argument could be convincing to readers due to the credibility of Cook's position (Go, Jung and Wu, 2014). Since the inclusion of a source in an article contributes to enhancing the source's chosen frame (Van Gorp, 2007), this shows that two advocates of the Important frame (Zuckerberg and Cook) have, again, been successful in influencing the news to exhibit their desired frame. This is further evidence to suggest frame advocates within the social institutions factor (Shoemaker and Reese, 2014) played a significant role in the framing of XR.

Another rhetorical framing device used to create the Important frame was associating XR with high-profile or well-known companies. For instance, one *MailOnline* article noted: "Big companies such as Apple, Facebook, Sony and Samsung have big stakes in the emerging sector" (Shoffman, 2014) and another highlighted that "HTC, Lenovo, Asus, and HP" are working with Microsoft on HoloLens (Beall, 2016). For these successful and established companies to be mentioned in relation to XR suggests the technology is significant because they have deemed it worthy of investment. In a similar way, it was also mentioned that Google was one of the major investors in Magic Leap. One *Guardian* article stated: "The investors are also of an unusually high calibre, including Google and the semiconductor magnate, Qualcomm" (Cumming, 2014). Again, the involvement of these well-known and successful companies have been noted to highlight the significance of the technology. Moreover, by using the modifier "high calibre" when referring to the investors, even readers who are unfamiliar with these companies will be aware that they are well-established in their industries, thus accentuating the same importance of Magic Leap.

Nordfors states that the reputation of an innovation depends greatly "on the reputation of the innovator, especially the innovator's reputation of innovating" (2009: 15). Although Facebook/Meta does not have a highly positive reputation, particularly in relation to the privacy of its users (Johnson, Egelman and Bellovin, 2012), it certainly has a strong reputation as an innovator. Indeed, the social media platform grew from approximately one million monthly users in 2004 (Sedghi, 2014) to approximately two billion active users by the end of 2022, with an annual revenue of $116.6 billion in the same year (Meta, 2023). Therefore, highlighting Facebook's involvement with VR works towards improving the reputation of Oculus and the wider XR industry. The same can be said for the other successful technological innovators mentioned in the previous paragraph. As a result, these associations put particular emphasis on the Important frame within XR news.

Additionally, referencing the impact XR can have also contributed to creating the Important frame. For example, an article in the *Guardian* stated that: "Many of these filmmakers and journalists see VR as a way to cut through viewers' complacency about disaster or war stories" (Dredge, 2016a). This relates to the idea that VR can be the "ultimate empathy machine" (Milk, 2015), as discussed in Chapter 2. Filmmaker Chris Milk argues that the immersive capabilities of VR mean users are able to feel more empathy for certain people or groups by experiencing the world from their perspective in VR. This news article implies that VR can encourage the public to act by increasing empathy. It further highlights this with a quotation from Milk himself: "What you're talking about at some point is more than a medium, but is fundamentally an alternative level of human consciousness" (Dredge, 2016a). Here, the significance of VR is implied to be strong because it can alter human consciousness, thus appropriating the Important frame. This framing device highlights the observability of XR, or "the degree to which the results of an innovation are visible to others" (Rogers, 2003: 16). The greater this visibility is, the more likely consumers are to adopt an innovation (Rogers,

2003). Thus, emphasising the impact of XR to construct the Important frame could aid the diffusion of this technology.

Aside from the small number of words that appeared that could counter the Important frame, no other framing devices were uncovered to counter it. Bednarek and Caple state that "evaluations of Unimportance [...] are rare in news discourse, presumably because they decrease news value" (2012: 141). This certainly seems to be the case in news coverage of XR. Instead, it appears that the news values of prominence (Bednarek and Caple, 2012) and magnitude (Harcup and O'Neill, 2017) have caused journalists to frame XR as Important. The prominence news value can be observed in the emphasis on the large successful companies involved in XR, while the magnitude news value appears to have been considered because the impact of the technology is shown to be significant. Since news values are related to routine practices (Shoemaker and Reese, 2014), it appears that this factor has impacted the creation of the Important frame.

As well as improving the perception and observability of XR, the Important frame has significant consequences for XR diffusion. Maisch et al. argue that uncertainty over the importance of an innovation "can lead to resentment and aversion" of that product (2011: 4). However, by highlighting the importance of XR technology, the news media have avoided creating resentment or aversion and have instead reassured the public of XR's significant role in society. This could then lead to an increased likelihood of XR being adopted. In other words, the appearance of the Important frame is further evidence to indicate that the news coverage supports XR diffusion. Yet again, the Important frame contributes to the news acting as a promotional tool for XR and supporting the commercial interests of those selling the devices. Indeed, it shows that favourable framing persists even when the same frame did not appear in the XR marketing, further promoting this technology to the public. Just as lifestyle journalism has been described as an extension of marketing (English and Fleischman, 2019; Kristensen, Hellman and Riegert, 2019), the same seems to be the case in XR news.

Successful

When an innovation is in the early stages of the diffusion process (as XR was during the sample period of this study) it is not known whether it will be successful or not. Despite this, one frame that emerged from XR news articles was Successful. This involved presenting XR as a technology that is, or will be, a success. The current section examines the framing devices used to represent XR as Successful.

Firstly, across the whole sample, 17 percent of articles used words in the "successful" category, indicating the use of the Successful frame. In comparison, just 5 percent of articles included terms from the "unsuccessful" category. Thus, words relating to the success of XR were used in three times more articles than those implying XR is unsuccessful. Moreover, every news outlet used words in the "successful" group in significantly larger portions of their articles than terms in the "unsuccessful" category. Similarly, terms in the "successful" group were used more than "unsuccessful" words in every year studied. This shows that the news articles have consistently favoured a positive framing of XR over a critical one, regardless of news outlet or year. Since the news is the public's main source of information about emerging technologies (Whitton and Maclure, 2015; Williams, 2003), this could lead to positive perceptions of XR in terms of its success.

Although words in the "successful" category always dominated, there were some variations in how much these words were used per news outlet. Terms in the "successful" category were mentioned in a similar portion of articles in the *Sun* and *MailOnline* (13 and 14 percent respectively). Likewise, both of these news outlets used words in the "unsuccessful" category in a similarly low portion of articles (3 percent in each). Therefore, these news outlets do not appear to differ in terms of how often they apply the Successful frame to XR. On the other hand, the *Guardian* had the highest portion of articles including words from both of these categories, with a relatively large 25 percent using terms from the "successful" category and 11 percent including words from the "unsuccessful" group. This shows that the *Guardian* discussed the success (whether successful or unsuccessful) of XR more than the other news outlets. Additionally, the *Guardian* used words in the "successful" category over twice as often as those in the "unsuccessful" category and substantially more than the other news outlets. Thus, the Successful frame had significant prominence in the *Guardian*. Since audiences typically assign more credibility to quality news outlets such as the *Guardian* than they do tabloids (Frewer, Scholderer and Bredahl, 2003), the potency of the Successful frame in the *Guardian* could have a meaningful impact on readers. Furthermore, this variation shows that the media organisation factor (Shoemaker and Reese, 2014) has impacted the strength of the Successful frame, although not to the extent that it is present in some outlets and not others.

Additionally, while there were no substantial differences between VR and AR/MR articles' use of words in these categories, examining this data across the years of the sample period provides notable insight. Although words in the "successful" category were used more than words in the "unsuccessful" group every year, the use of these words was not stable over time. In fact, there was a large increase in the use of words in the "successful" category from 2013 (6 percent) to 2014 (25 percent). As previously mentioned, 2014 was the year Mark Zuckerberg acquired Oculus, spurring much interest in XR. Therefore, this data indicates that Zuckerberg's involvement with XR contributed not only to increased media attention of XR (see Chapter 3), but an increase in the positive framing of XR. This coincides with the findings in relation to the Social frame discussed in the previous chapter.

In more detail, the use of specific words in each of these categories indicates that the success of XR was highlighted most often by claiming it has a large audience. This can be demonstrated by the fact that the words *mainstream* and *popular* were the most used in the "successful" category. In a similar way, portrayals of XR as mainstream were more common than describing it as niche. The term *niche* only appeared in 2 percent of articles, whereas *mainstream* appeared in 7 percent. Moreover, rhetorical framing devices were used to argue that XR will have a large audience. For instance, the *Guardian* wrote the following about VR:

> In the same way as the Nintendo Wii's motion-oriented gaming opened up the industry to new users, from children to grandparents and casual gamers everywhere, VR could have a similar impact. Ashforth says: "I've tried it with my kids, my mum, everyone loves it".
>
> (Gibbs, 2014)

Here, an analogy is used as a rhetorical framing device by relating VR to a previous technology. VR is said to have a potentially similar impact as the Nintendo Wii did in terms of attracting a wide audience. Frames work by "connecting the mental dots for the public" (Nisbet, 2010: 47). Thus, by relating VR to a device that is known for widening the videogame market, the persuasiveness of the Successful frame is enhanced. To add to this, the statement is supported by a technical framing device in the form of a quote from a Sony employee (Ashforth) who states that people of all different age groups have enjoyed VR. Again, this implies VR will appeal to a wide audience. Moreover, the source was defined by the journalist as a senior game designer. The labels (or designators) applied to sources can indicate the level of authoritativeness of a statement (Bell, 1991; Pan and Kosicki, 1993). Referring to the source as a *senior* game designer presents him as established in the industry, thus affording his statement more credibility. As a result, the Successful frame is emphasised.

Aside from highlighting the audience size, the news articles also presented XR as established to construct the Successful frame. For instance, an article in the *Guardian* asked: "When your grandkids ask where you were [when] virtual reality took off, what will you say?" (Dredge, 2016b). This question implies that there is no doubt that VR will be successful and it will become such a major part of life that future generations ("grandkids") will want to know about the moment it became established. Additionally, a *MailOnline* article included the following statements:

> The world of virtual reality is hotting up [...] VR is one of the biggest trends in technology at the moment, with dozens of firms jumping on the bandwagon and developing VR headsets.
>
> (Bell, 2016)

Noting VR is "one of the biggest trends" and that many companies are developing headsets makes the industry appear very current and established. Additionally, this is not just mentioned in relation to VR. Another article from the *Guardian* argued that Google Glass is known by everyone apart from "those who have been vacationing on Mars" (Naughton, 2013), again suggesting this is an established product that is common knowledge. These depictions act as framing devices for the Successful frame. Presenting XR as established could reduce uncertainty about the technology, leading readers to be more accepting of it (Rogers, 2003) and thus increasing the likelihood of adoption.

Moreover, data regarding product sales and XR revenue were used as technical framing devices in the news articles to construct the Successful frame. This was often alongside rhetorical framing devices in the way that these statistics were evaluated by the journalist. One of the earliest examples of this appeared in an article about Oculus Rift's Kickstarter campaign from the *Guardian*. It stated: "Oculus raised $2.4 m for its Rift headset in September 2012, exceeding its initial fundraising goal by ten times. It remains one of the largest ever Kickstarter campaigns" (Hern, 2014). Highlighting that the campaign exceeded its goal by a very large amount creates the impression that the device is very popular, thus depicting the Successful frame. This is further emphasised by noting it is one of the largest Kickstarter campaigns.

Similar sentiments were also bolstered by sources. For instance, a *MailOnline* article stated: "Goldman Sachs has predicted VR and augmented realty as a segment will be worth $80 billion (£56 billion) by 2025, which is around the same size as the desktop computer market today" (Griffiths, 2016). In this sentence, the VR/AR industry is predicted be worth the same amount of money as a very established piece of technology (the desktop computer) by 2025. This comparison suggests VR and AR will be very financially successful. In a similar way to comparing VR to the Wii above, associating XR with the desktop computer allows readers to more easily make connections between new and existing information, which is important in making a frame salient (Nisbet, 2010). Additionally, the *Guardian* implied success with a quote from a device creator:

> Mike Jazayeri, director of product management at Google VR, says he is pleasantly surprised by the success: "We never imagined the momentum it has had. Immediately we got a lot of interest from content creators, brands, developers – and a year later more than a million Cardboards have shipped and there's hundreds of apps".
>
> (Tucker, 2015)

The success of Google Cardboard in particular is highlighted in this quote by mentioning the high volume of interest, content and sales of the device. Both of these examples also employ the technical framing device of an established source, which increases the persuasiveness of the frame (van Dijk, 1988; Go, Jung and Wu, 2014). Furthermore, the second example shows that the creator of a VR product has also acted as a frame advocate (relating to the social institutions factor of the hierarchy of influences model (Shoemaker and

Reese, 2014)) for the Successful frame, again emphasising the impact of such voices on the portrayal of XR.

In relation to actual product sales, several articles highlighted the fact that XR devices sold out quickly. Gear VR was said to have "sold out within hours of going on sale" (Volpicelli, 2015) and, more extremely, "the first wave [of Oculus Rifts] sold out on the firm's website in seconds" (Prigg, 2016b). Moreover, the headline of one *MailOnline* article read: "HTC reveals it sold 15,000 Vive VR headsets in the first 10 MINUTES of going on sale" (Woollaston, 2016a). Each of these articles emphasise the popularity, and thus success, of the devices. Additionally, the fact that this final example appeared in the headline of the article highlights the prominence of the frame (Pan and Kosicki, 1993). It is emphasised further with the use of capitals ("10 MINUTES") to imply that it is extraordinary to sell 15,000 headsets in that amount of time. In all, the use of numbers in each of these examples act as "[s]ignals that indicate precision and exactness" which can increase the persuasiveness of statements (van Dijk, 1988: 84). Therefore, constructing the Successful frame in such a way gives it particular salience.

Lastly, an exemplar was used as a framing device in the form of comparing the current wave of XR to the first wave of XR. As mentioned in Chapter 2, the 1990s saw the first attempt at consumer VR, though it was not commercially successful (Dixon, 2016). The apparent success of the new generation of XR products was sometimes emphasised by comparing it to this historical failure of VR products. One article from the *Guardian* began by stating:

> The first wave of VR headsets flopped, but soon the Oculus Rift, HTC Vive and PlayStation VR will go on sale – and they're going to be much, much better.
>
> (Hern, 2015)

This introductory paragraph highlights the failure of the "first wave of VR", but contrasts this with some of the new headsets being released in 2016. Though this does not directly state that the new products will be successful, they are said to be of a much higher quality and therefore the chance of success is insinuated to be higher than it was previously.

Similarly, a *MailOnline* article cited Magic Leap CEO Rory Abovitz saying "virtual reality and augmented reality are old terms, with a largely disappointing history", followed by the quote: "We have the term 'cinematic reality' because we are disassociated with those things" (Griffiths, 2014a). In other words, the Magic Leap product will be different to the historically "disappointing" attempts at XR. Including a statement from a source in a news article "makes a positive contribution in the evocation of a frame" (Van Gorp, 2010: 103). The journalist's choice to include Rory Abovitz as a source contributes to framing XR the way Abovitz did: Successful. Again, an XR company owner has been used as a technical framing device for the

Successful frame, showing the prominence of these sources as advocates at the social institutions level (Shoemaker and Reese, 2014).

While these examples demonstrate that various framing devices were used to construct the Successful frame, it is also important to acknowledge attempts within the news articles to counter this frame. When XR was portrayed as unsuccessful, this usually occurred in relation to Google Glass. The device was described as "an expensive flop" (Woollaston, 2015) by the *MailOnline* and the *Guardian* noted "the company [Google] has given up on trying to sell them [Google Glass] as a mainstream idea" (Dredge, 2016b). Indeed, an entire *MailOnline* article was dedicated to discussing the failure of the device, headlined: "Is Google Glass a flop? Developers – and customers – are ditching the smart spectacles in favour of Oculus Rift" (Griffiths, 2014b). The article noted that nine out of 16 developers who were working on Google Glass applications had cancelled their development. Moreover, the article stated that "its prospects of becoming a consumer hit in the near future are slim" (Griffiths, 2014b). Altogether, Google Glass is portrayed as unlikely to be successful. This shows that the technological characteristics of the devices have had more impact on the frames used than was obvious from the quantitative data discussed at the start of this section.

Nevertheless, the numerical data above, combined with the multiple framing devices used to construct the Successful frame, show that the news articles favoured this positive representation of XR over a negative perspective. The presence of the Successful frame is significant because it reduces the uncertainty surrounding XR. During the innovation-decision process, individuals aim to reduce uncertainty about an innovation (Rogers, 2003). The Successful frame arguably reduces this uncertainty because it presents XR as something that has already been established and is of a high enough quality to achieve success. In this way, the Successful frame could promote the diffusion of XR. Moreover, the perception of a new technology has a significant impact on its acceptance (Buenaflor and Kim, 2013). Thus, for the news articles to present XR in a favourable light using the Successful frame could generate positive perceptions of XR. As a result, audiences may be more likely to adopt the technology. This shows that the news aids the promotion of XR by presenting it as Successful even when it was not yet known whether it would be. Such news coverage aligns with the commercial agendas of XR companies by supporting adoption and diffusion.

Affordable

While the current generation of XR products cost much less than they did during the first wave of XR in the 1990s (Fuchs et al., 2017; Steinicke, 2016), the different products still vary in price. For instance, a Google Cardboard VR headset costs approximately $6, whereas the Microsoft HoloLens MR device costs $3,000 (Greengard, 2019). Valuing between these figures, the headset that was mentioned the most in the news articles, Oculus Rift, cost $599 when it launched in 2016 (Morris, 2016). Despite these variations in price, it was found that one of the frames applied to XR in the news articles was Affordable. This involved presenting XR products as reasonably priced instead of overpriced or expensive. The current section analyses the framing devices used to construct the Affordable frame in XR news.

To begin, across the news outlets, 15 percent of articles used words from the "affordable" category. The preference for portraying XR this way is highlighted by comparing this figure with the number of articles using words that would counter an Affordable frame: terms in the "expensive" category appeared in 8 percent of articles. Additionally, every news outlet used "affordable" words in more articles than they did "expensive" words. Still, there were some differences in the prevalence of these words depending on the news outlet. Mirroring the results regarding the Successful frame, the *Guardian* used words in the "affordable" category in the most articles (19 percent). The *MailOnline* used such words slightly less (15 percent). Both of these outlets used terms in the "expensive" category considerably less (8 percent of articles each). However, the *Sun* rarely mentioned words in either of these categories (7 percent of articles included "affordable" words and "expensive" words appeared in 3 percent). This suggests that the *Sun* did not focus considerably on the price of XR products, although the outlet was still more likely to portray them as affordable rather than expensive. These figures show that the Affordable frame was used by the *Guardian* and *MailOnline* more so than the *Sun*. Therefore, the media organisation (factor three in the hierarchy of influences model (Shoemaker and Reese, 2014)) reporting on XR has had an impact on the strength of this frame, as was found to be the case for most other frames.

Additionally, this is one of the few frames that appeared to differ significantly between VR and AR/MR products. Words in the "affordable" category appeared in 16 percent of VR articles but only 3 percent of AR/MR articles. Moreover, words in the "expensive" category appeared in 8 percent of VR reports and 4 percent of articles about AR/MR. These figures suggest that evaluating the price of AR/MR products was much less common than it was for VR devices. Therefore, although the Affordable frame has been applied to VR articles, this does not seem to be the case in AR/MR coverage as these words were used so few times. Since the AR/MR devices this sample focused on cost substantially more than the VR products, this is not a surprising finding. For instance, of the devices mentioned in ten or more articles, the lowest priced AR/MR product was Google Glass at $1,500 (Greengard, 2019), whereas the most expensive VR product (HTC Vive) cost $799 (Burgess, 2016). However, it does show that, despite AR/MR products costing more than VR devices, the news outlets still have not portrayed them as expensive. Again, they have avoided critical representations of XR.

Furthermore, the use of words in the "affordable" category varied over time. They were most common in 2015 (22 percent), indicating that the

Affordable frame was most prominent in this year. However, in 2016 (the year that several major VR headsets were released to the public), the appearance of words in the "expensive" category peaked at 13 percent of articles. This suggests that there were the most attempts to counter the Affordable frame in this key year for the XR industry. According to Sääksjärvi and Morel (2010), one factor that could lead consumers to reject a technology is doubt over perceived value for money. Since 2016 was the year that many consumers would make their decision of whether to buy a dedicated VR device, the rise in words countering the Affordable frame could have increased their doubt about VR's value for money and thus reduced their willingness to buy the product. Nevertheless, it should be remembered that words in the "affordable" category were used more than those in the "expensive" category in every year of the sample, showing that the Affordable frame was favoured by journalists overall.

As the Affordable frame appears to have been contested more so than others, it is worth examining the framing devices used to both support and counter this frame. Aside from the use of specific words, the Affordable frame was observed in the news articles through speculation about the price of the products. For example, before the price of Oculus Rift had been announced, a *MailOnline* article was published with the headline: "Facebook's Oculus Rift Virtual Reality headset to cost just $200" (Prigg, 2014b). Using the modifier "just" implies this is a low amount. However, this headline is misleading, as the article itself stated that "it has been revealed it could cost as little as $200" (Prigg, 2014b). Though the headline implies this supposedly low price is certain, the body indicates that this is only a possibility. Moreover, further in to the article, the journalist detailed that: "[Oculus Rift] will be offered for around $200–$400, according to Oculus VR co-founder Nate Mitchell" (Prigg, 2014b). Here, in a less prominent part of the article, a higher price is introduced. As the headline and lead of an article are the most powerful in creating a certain frame (Pan and Kosicki, 1993), this shows that the journalist chose to emphasise the lowest figure mentioned by the device creator, thus framing XR as Affordable.

In a different way, comparisons were also used as a rhetorical framing device to construct this frame. For example, a very early article about Google Glass published in 2012 stated that the device could cost "less than £380 – making it cheaper than Apple's iPhone" (Narain, 2012). Whereas the journalist could have used a more neutral word, such as "approximately £380", their use of the word "less" implies £380 is a low amount. This is emphasised by noting it is "cheaper" than a product bought by millions of consumers – the iPhone. Certainly, in the year this article was published (2012) Apple shipped 136.8 million iPhone units (*AppleInsider*, 2013). Since the iPhone is a product that is very popular, it seems many consumers consider this a reasonable price for a phone. For Google Glass to cost even less implies the price is just as reasonable and perhaps even more so since the device has also been framed as Advanced and High-Quality (see Chapter 5), thus depicting the Affordable frame.

Whereas news articles used the Affordable frame when the prices of products were not yet known, some articles exaggerated the price of Oculus Rift after it had been announced. For example, an article in the *Guardian* stated: "Oculus will sell for about $1,500 (although this includes a powerful PC to drive its graphics)" (Tucker, 2015). The *MailOnline* used a very similar statement in one of its headlines: "Facebook's Oculus Rift headset will cost $1,500 (including the new computer you'll probably need to power it)" (Prigg, 2015b), making it even more salient by including it in the most significant part of the news article (Pan and Kosicki, 1993). Although these articles mention that this price includes the new PC needed to use the headset, it is not specified how much of that figure is for the headset itself or the computer. Therefore, it creates the overall impression that the product is expensive, countering the Affordable frame.

Certainly, when attempts were made to counter the Affordable frame by suggesting XR was expensive, this usually focused on the external components needed for the experience; namely, the PCs already mentioned. Some examples of this are as follows:

> The headset will also require an expensive, high-powered PC to run VR applications
>
> (Wong, 2016)

> you'll probably need an expensive new PC to run it
>
> (Prigg, 2016a)

> headsets to view VR video can cost more than $1,000 once you include a high-end personal computer with fast-enough graphics
>
> (Associated Press, 2016)

Alternatively, some mentions of price were less specific, with one *MailOnline* article stating that the "main problem" with VR "is its price" (Woollaston, 2014). There were also instances where the headsets themselves were represented as expensive, such as HoloLens' price of £2,719 being described as "gargantuan" by Parkin (2016a) in the *Guardian*. Therefore, it can be seen that there were some attempts to oppose the Affordable frame within the news coverage.

Relatedly, comparisons were used as framing devices to present certain products as more affordable than others. This was identified in relation to PlayStation VR. The headset was framed as Affordable because it does not need to be connected to a PC to work. For example, the headline of a *MailOnline* article stated: "Sony's PlayStation VR to undercut Oculus and HTC: Headset will cost $399 and you WON'T need an expensive new PC to use it" (Prigg, 2016c). Additionally, a journalist for the *Guardian* wrote:

Sony's virtual reality headset, the PSVR, will launch globally in October, for the comparatively low price of £349.

It's unusual for a peripheral that costs more than its host game console to be considered a bargain, but virtual reality is proving to be a pricey frontier for early adopters. HTC's Vive will retail for $799/£689, while Facebook's Oculus Rift, which will launch in April, costs $599/£499, a significant amount when you consider the additional cost of the formidable PCs required to run the hardware competently.

(Parkin, 2016b)

Firstly, PlayStation VR's price is described as "comparatively low", which implies that, although it might not usually be considered cheap, it is reasonable when measured against the cost of other VR devices. This is emphasised in the beginning of the second paragraph when the article mentions the headset ("peripheral") costs more than the console needed to use it, though is still considered a bargain because of the prices of the other products. In this example, HTC Vive and Oculus Rift are presented as expensive in comparison to PlayStation VR. The high cost of HTC Vive and Oculus Rift is further highlighted by the mention of the "additional cost of the formidable PCs" needed to experience VR. In both of these examples, PlayStation VR is framed as Affordable, whereas HTC Vive and Oculus Rift are presented as expensive in comparison because of the PCs needed to use them. Therefore, it appears that the technological characteristics of specific products have impacted the strength of the Affordable frame.

In sum, although the Affordable frame was not as prominent as others due to there being more attempts to counter it, the news articles certainly did not favour a critical view when it came to the price of this technology. The perception of an innovation's value for money can have a significant impact on a consumer's decision of whether or not to adopt it (Sääksjärvi and Morel, 2010). Similarly, perceived fee is one of the two main sacrifices in Kim, Chan and Gupta's (2007) value-based adoption model (VAM). According to the authors, the higher the perceived fee, the less likely consumers are to adopt a technology. The fact that the Affordable frame has been the most salient (as opposed to a frame portraying XR as expensive) could lead readers to perceive the fee of XR products to be reasonable. Thus, the Affordable frame could promote the diffusion of XR.

Like the Comfortable and Easy to Use frames discussed in the previous chapter, any discussion of price indicates that journalists assume their readers to be interested in purchasing one of these products since the price would be irrelevant otherwise. Indeed, the mention of price was one of the factors considered by Arik and Çağlar (2005) in their analysis of consumption messages in Turkish lifestyle journalism. It is reassuring that the Affordable frame was contested more than others, but it was still the most dominant in comparison to its counterpart (i.e. expensive). As in Arik and Çağlar's study, the discussion of price indicates a discourse of consumption in XR news.

Since this price is positively evaluated in content that is presented as news, this benefits the companies aiming to sell these products.

Much-Anticipated

The final frame to be discussed is Much-Anticipated. The use of this frame emphasised excitement for XR that could then generate hype. Unlike the other frames in the chapter, this section does not examine quantitative data regarding words that could counter this frame because it does not have a clearly articulated opposite. Additionally, no attempts at countering this frame could be found within the news articles through qualitative analysis either. Therefore, this section examines the appearance of the Much-Anticipated frame, including its prevalence and which framing devices were used to construct it.

The strength of this frame can be clearly demonstrated by how often words in the "much-anticipated" category were used. Such terms appeared 480 times in 27 percent of articles overall. That is to say, out of all the word categories for specific frames, "much-anticipated" terms were the third most common. Moreover, out of all individual search terms, the stem *excit** (e.g. exciting, excitement) was used in the second largest portion of news articles, appearing in 18 percent. This indicates that the Much-Anticipated frame was particularly strong. Similarly, despite more VR products being commercially released during the sample period of this study than AR/MR devices, there was little difference in the use of these words depending on XR type. Thus, the news articles presented both types of XR as Much-Anticipated.

However, there was some variation per news outlet in how often these words appeared. Out of all frame-based categories, the *Guardian* used "much-anticipated" words the second most and more than any other news outlet (33 percent). Slightly less than the *Guardian*, words in the "much-anticipated" category were the third most used in the *MailOnline* (26 percent). Alternatively, the *Sun* was the least likely to use words from the "much-anticipated" category, with them appearing in 11 percent of articles from this outlet. In fact, words relating to four other frame categories were used more in the *Sun* than those referring to the Much-Anticipated frame. This data indicates that the strength of the Much-Anticipated frame varied depending on the media organisation factor (Shoemaker and Reese, 2014), although it was still present in every news outlet to some extent.

Importantly, there was some variation in the use of words in this category over time. Although words in the "much-anticipated" category were at their lowest point in 2012 (17 percent), this rose dramatically in 2013 to 39 percent where it reached its peak. In other words, the Much-Anticipated frame appears to have been the strongest the year after the second wave of XR began. From 2014 to 2016, terms in this category remained fairly consistent, ranging from 29 to 31 percent. However, this dropped to 18 percent in 2017, the year after several major VR products were released to consumers. This data suggests that the news articles attempted to increase the hype and excitement for XR leading up to the release of several products in 2016. Previous studies have suggested that consumer anticipation increases the chance a new product will be successful (Lee and O'Connor, 2003; Schatzel and Calantone, 2006; Vichiengior, Ackerman and Palmer, 2019). Therefore, the fact that the Much-Anticipated frame was prevalent in the years leading up to the release of many XR products could have supported its adoption.

In addition to the use of these specific words, rhetorical framing devices were used to construct the Much-Anticipated frame. One technique involved using modifiers to emphasise the anticipation surrounding XR. Some examples of this are shown below:

Sony's highly anticipated Project Morpheus

(Hern and Gibbs, 2014)

the highly-anticipated gaming gadget

(Griffiths, 2016)

the much anticipated Oculus Rift virtual reality headset

(Prigg, 2014c)

the eagerly awaited Touch controller

(Prigg, 2016d)

eagerly anticipated Vive virtual reality system

(Griffiths and Woollaston, 2016)

The use of the modifiers "highly", "much" and "eagerly" in these sentences implies there is substantial excitement surrounding XR, thus framing it as Much-Anticipated. Similarly, an article from the *Guardian* claimed: "it's VR that has everyone excited" (Arthur, 2015). For "everyone" to be excited about VR implies the technology must be worthy of this excitement, thus potentially generating more hype and anticipation for the technology. In another article, this was combined with an exemplar to further emphasise the excitement surrounding XR. The opening line of a *MailOnline* article described Oculus Rift as "one of the most anticipated gadgets since the iPhone" (Prigg, 2015a). Using the iPhone as an exemplar implies that the same level of interest surrounds Oculus Rift as did the popular smartphone. Considering the strong success of the iPhone, this suggests that there is extreme excitement surrounding Oculus Rift. Emphasis and exclusion are major parts of framing an issue or topic (Gitlin, 1980; Hallahan, 1999; de Vreese, 2010). These examples demonstrate just how much the anticipation over XR was emphasised in the news articles, thus increasing the salience of the Much-Anticipated frame.

Furthermore, the use of the word "finally" works as a rhetorical framing device in the articles to emphasise anticipation for XR. This is combined with technical framing devices to add prominence to these points. For instance, a *MailOnline* article headline claimed "Virtual reality is finally here" (Anwer, 2015). The use of the word "finally" suggests that much time has passed waiting for this technology. Thus, for VR to be "finally here" seems even more significant and worthy of excitement and hype. The fact that this point appeared in the headline of an article demonstrates the salience of this argument (Pan and Kosicki, 1993) and, in turn, the Much-Anticipated frame. Moreover, the *MailOnline* also wrote that "the Oculus Rift headset finally delivers on the long-awaited promise of virtual reality" (Prigg, 2014a). In this example, it is not just the device itself that is shown to be long-awaited, but VR in general, making it seem even more noteworthy. The rhetorical framing device works together with a technical framing device to increase the prominence of this point, since it appeared in a side-note of 16 *MailOnline* articles from 2014 to 2016. Again, *MailOnline*'s routine practice (factor four in the hierarchy of influences (Shoemaker and Reese, 2014)) to repeat sections of its articles has worked to emphasise the Much-Anticipated frame. This is likely an effect of the capitalist social system (the highest level factor in the hierarchy of influences model) causing journalists to be under pressure to create news content quickly for commercial gain.

According to Newman, the "pages of the specialist gaming press brim over with anticipation, communicating palpable longing and desire for the next game" (2012: 60, quoted in Vollans et al., 2017: 1). The existence of the Much-Anticipated frame in XR news suggests that generalist news coverage of XR is similar to that of the specialist gaming press mentioned by Newman. However, whereas games journalists typically present their articles as reviews (Foxman and Nieborg, 2016), XR news was most commonly presented as traditional news (see Chapter 3). This means audiences would have different expectations when reading these types of articles, being under the impression that what is presented as "news" is based on research and unbiased facts (Pan and Kosicki, 1993). This could make these frames more persuasive, effectively disguising promotion as news. Again, it appears XR news has been marketised (Fairclough, 1993), blurring the distinction between news and promotional content.

Moreover, the combination of this frame with others that positively evaluate XR (as discussed above) could lead to considerable hype and excitement over the technology. On the one hand, hype leads to "high rising expectations about the potential of [an] innovation" (Ruef and Markard, 2010: 317), which could support its adoption (Hedman and Gimpel, 2010). On the other hand, hype is usually followed by disappointment. Ruef and Markard state that the "subsequent drop of attention and a disappointment of the hyped expectations may have negative effects on the innovation process" (2010: 317–318). Therefore, although the Much-Anticipated frame could initially support the diffusion of XR, it may not have a long lasting positive effect. Nevertheless, at least in the initial stages, this frame works to support the promotion of XR, aligning with the goals of XR companies.

Positive Framing of Extended Reality

Up to now, this book has mostly examined specific frames in the news coverage of XR, finding that the frames which present XR in a positive light are favoured over any that might criticise the technology. This is a strong indication that coverage is more positive than negative. To further understand the overall framing of XR news coverage (and indeed whether this contributes to the diffusion of XR), it is useful to analyse the general tone of the articles regardless of which specific frame is being used. To investigate the overall tone, the study recorded the use of positive and negative words within the articles, as well as any words relating to concerns and ailments surrounding XR. The qualitative analysis also explored which framing devices had been used to present XR in a positive or negative light. Thus, instead of focusing on a specific frame, this section examines how the general tone of XR news coverage has contributed to the overall framing of the technology.

Positive and Negative Discourse

At a broad level, the use of positive and negative words within the news articles can indicate the overall tone of the coverage. In XR news, words in the "positive" category were used in 53 percent of articles, whereas words in the "negative" category appeared in 29 percent of articles (see Table 7.1). Additionally, individual uses of "positive" words appeared over twice as often as individual uses of "negative" words. These figures are further indication that articles were more likely to frame XR positively than negatively, though negative words were not completely absent. This remained consistent throughout the years studied, with every year seeing a substantially larger portion of articles using words in the "positive" category than in the "negative" category (see Figure 7.1). Examining the use of terms in both of these groups across time shows that the trajectories they followed were very similar. Therefore, rather than indicating how coverage became more or less positive/negative over time, this data suggests that articles were simply more likely to evaluate XR in certain years than others. Furthermore, an important finding is that the use of "positive" words peaked in 2016, at 57 percent. This shows that the year several dedicated VR devices were released to consumers saw the most positive evaluations of XR. In this year, potential early adopters would have been in the decision stage of the innovation-decision process (Rogers, 2003). Therefore, the fact that "positive" words dominated could mean that the news has promoted XR diffusion by presenting the technology in a favourable light when this decision was being made.

Considering the differences between articles focusing on VR or AR/MR highlights a similar point. Slightly more "positive" words were used in VR

Table 7.1 Appearance of terms in the "positive" and "negative" categories per news outlet

Category	Sun		Guardian		MailOnline		Overall	
	Uses	Percent	Uses	Percent	Uses	Percent	Uses	Percent
Positive	60	54.10	459	63.31	785	48.80	1304	52.81
Negative	28	18.03	236	45.16	328	23.35	592	28.56

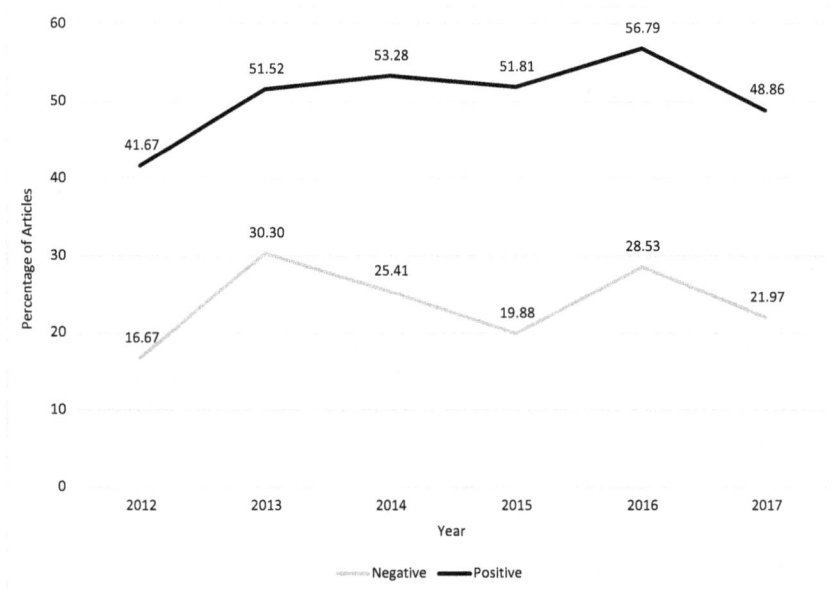

Figure 7.1 Percentage of articles per year with at least one term from the "positive" and "negative" categories.

articles than AR/MR articles (55 percent compared to 42 percent). However, the same trend appeared regarding "negative" words, with these appearing in 24 percent of VR articles and 21 percent of AR/MR news reports. Therefore, the technological characteristics of XR do not seem to have impacted whether they would be framed more positively or negatively, but simply how often they were evaluated in the news.

On the other hand, there were some differences in the use of "positive" and "negative" words between the news outlets in the sample. Whereas every news outlet used "positive" terms more than "negative" terms, the difference between the two categories varied. The *MailOnline* was the only news outlet not to use "positive" words in at least half of its articles (49 percent).

However, the publication used "positive" words in double the amount of articles that it used "negative" words, showing a positive tone was still favoured. Additionally, the *Sun* was the least likely to use "negative" words, with them appearing in 18 percent of articles. Three times more *Sun* articles included "positive" words compared to "negative" words. Therefore, this news outlet in particular appears to have emphasised positive coverage of XR, despite not reporting much on the topic. On the other hand, the *Guardian* used words in the "positive" category the most (63 percent), though it also used words in the "negative" group more than any other outlet (45 percent). This implies that the *Guardian* was most likely out of the three to use evaluative words, which is surprising considering the quality news outlet in this sample would be most expected to adhere to the journalistic norm of objectivity (Bastos, 2019). Nevertheless, the *Guardian* has been more balanced in its news coverage than the other outlets, with the smallest difference between the use of "positive" and "negative" words compared to the *Sun* and *MailOnline*. This indicates the publication may have been more analytical about XR than the others, in line with the norms of quality news outlets. Still, "positive" words were used more by all publications, demonstrating that framing XR in a positive light was a trend shared across the sample. Since moral panics involve exaggerated fear and negativity (Cohen, 2002; Hall et al., 1978), this is further evidence to support the claim that a moral panic does not exist in XR news.

In addition to considering how often words in these categories appeared, examining which specific words were used the most within these groups sheds more light on the strength of positive and negative evaluations. Entman states that "content analysis informed by a theory of framing would avoid treating all negative or positive terms or utterances as equally salient and influential" (1993: 57). Indeed, Bednarek and Caple argue that "[e]valuations of Emotivity [or tone] are expressed by a range of linguistic items that vary enormously in their evaluative force and are situated on a cline ranging from more or less positive to more or less negative" (2012: 144). With this in mind, it was found that several words with strongly positive connotations were used in XR news coverage. For instance, although *good* was common (appearing in 8 percent of articles), this word does not imply something is exceptionally positive, but is a rather mediocre evaluation. On the other hand, out of the 53 words in the "positive" category, the terms *great*, *amaz** (e.g. amazing, amazement) and *incredible* were among the ten most used. These words have stronger positive connotations and thus present XR in an even more positive light than *good*. In addition, the terms *best* and *perfect** also appeared in the top ten "positive" words. These terms are highly positive since they imply that something could not possibly be any better than it already is.

Alternatively, strongly negative words were rarely used in the news coverage. As the opposite to *good*, the term *bad* was used just 20 times in 2 percent of articles – less than any of the "positive" words mentioned in the previous paragraph. Additionally, strongly negative words were very rarely used. For instance, the term *awful* did not appear at all and *terrible* only appeared three times. This shows that positive evaluations were much more salient in XR news coverage than negative evaluations. Furthermore, these results indicate that, as well as the use of specific frames presenting XR positively, the overall framing was also positive. Thus, instead of creating a moral panic around XR, this coverage encourages readers to develop a favourable view of XR. The perception of a new technology has a significant impact on its acceptance (Buenaflor and Kim, 2013). Therefore, the positive tone of the articles could support the diffusion of XR.

Beyond word usage, positive imagery was also used to add to the positive representation of XR. This involved including pictures of XR users with looks of happiness or wonderment as they interacted with the devices. For instance, images showed users smiling while wearing headsets such as Oculus Rift (Dredge, 2014a), Google Glass (Woollaston, 2013) and Google Cardboard (Woollaston, 2016b). The inclusion of such images creates the impression that these products offer enjoyable experiences able to generate positive emotions. In turn, this presents XR in a favourable light. Other articles depicted users with expressions of disbelief and wonder when using the products, featuring Google Glass (Landi, 2017) and Oculus Rift (Dredge, 2014b, 2016a). In these images, the open mouths of users imply that they are impressed or awestruck by what they are seeing. While some articles used generic stock images to create this effect (e.g. Dredge, 2014b), an article in the *Sun* (Landi, 2017) included an example of a well-known individual reacting in this way: the UK's former Deputy Prime Minister Nick Clegg. Wearing Google Glass, Clegg's open-mouthed expression again indicates that he is impressed with the experience. Whereas political figures are often appropriated by the media to generate a moral panic (Hall et al., 1978), including in Sørensen's (2012) study of the videogame moral panic, the opposite appears to have happened here. Showing previous or current world leaders to be engaging with an XR product implies it must have reached a high level of significance within the technology industry. Furthermore, Clegg's expression of shock and wonder, combined with his position as a political elite, endorses the product and further supports the idea that it is impressive. Indeed, "images exert a more powerful influence on memory and perceptions than text" (Coleman, 2010: 243). Thus, using these images could be highly effective in presenting XR in a favourable light.

Attention to Concerns and Ailments

Aside from positive and negative framing devices, examining how often concerns were mentioned provides further insight into the overall tone of XR news coverage. Overall, 25 percent of articles mentioned words in the "concerns" category and 14 percent used words from the "ailments" category. Comparable to what was found regarding the "positive" and "negative" groups, the inclusion of words referring to concerns and ailments varied slightly per news outlet. Just as the *Guardian* was most likely to use words in the "negative" category, this publication also used words in the "ailments" and "concerns" categories the most (20 and 40 percent respectively). The *MailOnline* was the least likely to include words from the "concerns" category (19 percent) and used terms referring to "ailments" even less (12 percent). On the other hand, the *Sun* was least likely to include words in the "ailments" category (10 percent) but mentioned words relating to "concerns" in 25 percent of articles. The differences in these figures suggest that the media organisation factor (Shoemaker and Reese, 2014) has impacted how often ailments and concerns were noted in XR news.

Further differences were observed between articles about VR in comparison to those about AR/MR. VR articles were much more likely to include words in the "ailments" category, with such terms appearing in 16 percent of VR articles and just 3 percent of AR/MR articles. Since the most used words in the "ailments" category focused on cybersickness and eyestrain, which is more associated with VR than AR/MR devices, this shows that the technological characteristics of XR have impacted how often ailments were mentioned. Alternatively, words in the "concerns" category were mentioned in a much similar portion of VR articles in comparison to AR/MR articles (24 and 21 percent respectively). Therefore, although ailments were mentioned more in VR articles, the type of XR being reported on does not appear to have impacted how often concerns were mentioned.

Additionally, there are some notable points to be made about the use of these words across the years of the sample period. Words in the "concerns" category peaked in 2013 (42 percent) with a smaller peak in 2016 (31 percent). This first peak shows that concerns about AR/MR were particularly frequent in 2013, since this technology was the focus in that year. Moreover, the smaller peak in 2016 is significant because this was the year that several VR products were released. This data also shows that words in the "ailments" category were used fairly frequently in the years leading up to these releases (2014–2016). Maisch et al. state that "[b]efore consumers are prepared to adopt an innovation, they have to be convinced that the use of the innovation will not entail negative effects or unacceptable risks" (2011: 3). Therefore, the high use of words in these categories during this period could hinder the diffusion of XR.

Although these figures might suggest that particular attention was paid to the concerns and ailments surrounding XR, this type of news coverage can be better put into perspective by referring back to the discussion in Chapter 3. It was noted that just 3 percent of articles made Concerns their main topic and only two articles had the main topic of Regulation. Therefore, although these words were mentioned, they were not salient enough to become the main focus of many articles.

Certainly, although some articles note these concerns and ailments, other articles actually mention the way XR can be used to overcome the same issues. For example, the stem *sick** (relating to cybersickness) was the most common ailment to be mentioned, appearing in 6 percent of articles overall. However, there were also some reports that mentioned an XR device created to *prevent* motion sickness while flying. One article headline claimed: "The end of air sickness? Virtual reality headsets could prevent nausea on bumpy flights and even tackle jet lag" (Gray, 2015). Similarly, 17 articles used the stem *isolat** (e.g. isolating, isolation) to show concerns over the solitary XR experience. However, in other articles, it was said that XR can be used to help people escape isolation, including astronauts (Zolfagharifard, 2014), military personnel (Macdonald, 2016) and hospital patients and the elderly (Ferguson, 2016). Additionally, the idea of being isolated in a VR experience is not always portrayed as a negative. For example, a *Guardian* article described being able to isolate a patient using VR during an operation as a great advantage (Parkin, 2014). Moreover, isolation was also mentioned in a positive way in a *MailOnline* article in terms of flight passengers being able to isolate themselves from the other goings on in the plane to have a more pleasant journey (Kitching, 2016).

Considering this data, it is clear that although concerns and ailments were mentioned, they were certainly not a major focus of the news articles. Additionally, positive words were used much more often in the news coverage than negative words, complemented by positive imagery of XR users. These findings coincide with the results found by Allan, Anderson and Peterson (2010) on nanotechnologies, Cogan (2005) on the personal computer and Hetland (2012) on the internet, although differ from Whitton and Maclure's (2015) analysis of videogame news. Despite the focus on videogame applications, XR news coverage seems to have more similarities with these other emerging technologies (nanotechnology, computers and the internet) than videogames themselves. Furthermore, based on the definitions of moral panics discussed in Chapter 2, a moral panic surrounding XR has not been created by these three news outlets. Though words relating to concerns were mentioned, this was not a main focus of the articles. Thus, consensus (Goode and Ben-Yehuda, 1994) was not generated in relation to these negative aspects. Similarly, as opposed to technopanics that focus on regulation (Marwick, 2008), regulation was very rarely mentioned in news coverage of XR.

While moral panic coverage can be damaging for new technologies by resulting in unnecessarily strict legislation (Marwick, 2008), the news coverage of XR appears to have veered so far in the opposite direction that it is problematic in another way. The public depend on news media to make sense of new technologies and to generate public debate about their benefits, risks and social implications (Anderson et al., 2005; Dimopoulos and Koulaidis, 2002; Schäfer, 2017; Scheufele, 2013). The lack of critical coverage or attention to concerns mutes this discussion, which could lead to the absence of regulation. Certainly, even when writing this book in 2023, no XR-specific regulation exists. Instead of promoting discussion, the mostly positive tone of the articles could lead readers to form favourable views of the technology and, as a result, be more likely to adopt it (Buenaflor and Kim, 2013). Therefore, it appears that the news articles have supported the diffusion of XR in this way and, as a result, have prioritised the interests of the companies aiming to sell these products.

As noted in Chapter 3, the majority of XR articles were presented as news and therefore would be expected to contain facts rather than biased opinions as would be typical of, for example, review coverage. Since each frame and the overall framing positively evaluates the technology, the news discourse serves as a promotional tool for XR. The coverage appears to share similarities with lifestyle journalism in this way (which is often seen as an extension of marketing (English and Fleischman, 2019; Kristensen, Hellman and Riegert, 2019)), despite being presented as news. Therefore, XR news has the characteristics of evaluative journalism disguised as impartial news content, thus encouraging readers to accept such frames as fact. This coverage is now so far away from moral panic style discourse that it has the opposite problem – being overly positive with little attention paid to potential concerns or drawbacks that could spark public debate. With these results in mind, the final chapter of this book provides an overview of the findings of this study and considers the implications of these results. It also provides more detail about the model of frame categories and how it can be applied in future studies of emerging technologies.

References

AFP (2017) 'Microsoft takes aim at Facebook's Oculus with $299 'mixed reality' headset and motion controllers', *MailOnline*, 11 May. Available at: 8 February 2019).

Allan, S., Anderson, A. and Petersen, A. (2010) 'Framing Risk: Nanotechnologies in the News', *Journal of Risk Research*, 13(1), pp. 29–44. doi:10.1080/13669870903135847.

Anderson, A., Allan, S., Petersen, A. and Wilkinson, C. (2005) 'The Framing of Nanotechnologies in the British Newspaper Press', *Science Communication*, 27(2), pp. 200–220. doi:10.1177/1075547005281472.

Anwer, J. (2015) 'TECH SAVVY: virtual reality is finally here', *MailOnline*, 12 May. Available at: www.dailymail.co.uk/indiahome/indianews/article-3078733/TECH-SAVVY-Virtual-reality-finally-here.html (Accessed: 13 February 2019).

AppleInsider (2013) 'Apple's iPhone grew to 25.1% global market share in 2012', *AppleInsider*, 13 January. Available at: https://appleinsider.com/articles/13/01/25/apples-iphone-grew-to-251-global-market-share-in-2012 (Accessed: 9 October 2020).

Arik, M.B. and Çağlar, S. (2005) 'The Face of Consumption Society in the Press: Life Style Journalism', *Proceedings of the 3rd International Symposium Communication in the Millennium*, Chapel Hill, NC, 11–13 May. Available at: https://citeseerx.ist.psu.edu/viewdoc/download?doi=10.1.1.507.9260&rep=rep1&type=pdf

Arthur, C. (2015) 'The return of virtual reality: 'this is as big an opportunity as the internet", *The Guardian*, 28 May. Available at: www.theguardian.com/technology/2015/may/28/jonathan-waldern-return-virtual-reality-as-big-an-opportunity-as-internet (Accessed: 20 December 2018).

Associated Press (2016) 'VR on the cheap: how to watch without a headset', *MailOnline*, 15 March. Available at: www.dailymail.co.uk/wires/ap/article-3493706/VR-cheap-How-watch-without-headset.html (Accessed: 12 February 2019).

Bantimaroudis, P. and Ban, H. (2001) 'Covering the Crisis in Somalia: Framing Choices by *The New York Times* and *The Manchester Guardian*', in Reese, S.D., Gandy, O.H. and Grant, A.E. (eds.) *Framing Public Life*. London: Lawrence Erlbaum Associates, pp. 175–184.

Bastos, M.T. (2019) 'Tabloid Journalism', in Vos, T.P., Hanusch, F., Geertsema-Sligh, M., Sehl, A. and Dimitrakopoulou, D. (eds.) *The International Encyclopedia of Journalism Studies*. Malden, MA: Wiley. doi:10.1002/9781118841570.

Beall, A. (2016) 'Microsoft is letting others in on their virtual reality platform: Windows HoloLens now open for anyone to build devices', *MailOnline*, 1 June. Available at: www.dailymail.co.uk/sciencetech/article-3619273/Microsoft-wants-Windows-open-mixed-reality.html (Accessed: 12 February 2019).

Bednarek, M. and Caple, H. (2012) *News Discourse*. London: Continuum International Publishing Group.

Bell, A. (1991) *The Language of News Media*. Oxford: Blackwell.

Bell, L. (2016) 'Sony's PlayStation VR set to officially launch next week: invite-only event expected to reveal release date and price', *MailOnline*, 7 March. Available at: www.dailymail.co.uk/sciencetech/article-3480457/Sony-s-PlayStation-VR-set-officially-launch-week-Invite-event-expected-reveal-release-date-price.html (Accessed: 13 February 2019).

Buenaflor, C. and Kim, H. (2013) 'Six Human Factors to Acceptability of Wearable Computers', *International Journal of Multimedia and Ubiquitous Engineering*, 8(3), pp. 103–114.

Burgess, M. (2016) 'HTC Vive available now for £689, but shipping adds £57.60', *Wired*, 29 February. Available at: www.wired.co.uk/article/htc-vive-uk-price-release-date (Accessed: 9 October 2020).

Cogan, B. (2005) '"Framing Usefulness:" An Examination of Journalistic Coverage of the Personal Computer from 1982–1984', *Southern Journal of Communication*, 70(3), pp. 248–295. doi:10.1080/10417940509373330.

Cohen, S. (2002) *Folk Devils and Moral Panics*. 3rd edn. Oxon: Routledge.

Coleman, R. (2010) 'Framing the Pictures in Our Heads: Exploring the Framing and Agenda-Setting Effects of Visual Images', in D'Angelo, P. and Kuypers, J.A. (eds.) *Doing News Framing Analysis*. Oxon: Routledge, pp. 233–261.

Cumming, E. (2014) 'Magic Leap: startup promises a leap forward for virtual reality', *The Guardian*, 25 October. Available at: www.theguardian.com/technology/2014/oct/25/virtual-reality-leap-forward-google (Accessed: 14 December 2018).

de Vreese, C.H. (2010) 'Framing the Economy: Effects of Journalistic News Frames', in D'Angelo, P. and Kuypers, J.A. (eds.) *Doing News Framing Analysis*. Oxon: Routledge, pp. 187–214.

Dimopoulos, K. and Koulaidis, V. (2002) 'The Socio-Epistemic Constitution of Science and Technology in the Greek Press: An Analysis of its Presentation', *Public Understanding of Science*, 11, pp. 225–241.

Dixon, W.W. (2016) 'Slaves of Vision: The Virtual Reality World of Oculus Rift', *Quarterly Review of Film and Video*, 33(6), pp. 501–510. doi:10.1080/10509208.2016.1144018.

Dredge, S. (2014a) 'Oculus warns Sony to solve motion sickness before launching a VR headset', *The Guardian*, 4 November. Available at: www.theguardian.com/technology/2014/nov/04/oculus-sony-motion-sickness-virtual-reality (Accessed: 19 December 2018).

Dredge, S. (2014b) 'Oculus Rift — 10 reasons why all eyes are back on virtual reality', *The Guardian*, 31 March. Available at: www.theguardian.com/technology/2014/mar/31/oculus-rift-facebook-virtual-reality (Accessed: 19 December 2018).

Dredge, S. (2016a) 'Three really real questions about the future of virtual reality', *The Guardian*, 7 January. Available at: www.theguardian.com/technology/2016/jan/07/virtual-reality-future-oculus-rift-vr (Accessed: 20 December 2018).

Dredge, S. (2016b) 'The complete guide to virtual reality — everything you need to get started', *The Guardian*, 10 November. Available at: www.theguardian.com/technology/2016/nov/10/virtual-reality-guide-headsets-apps-games-vr (Accessed: 20 December 2018).

English, P. and Fleischman, D. (2019) 'Food for Thought in Restaurant Reviews', *Journalism Practice*, 13(1), pp. 90–104. doi:10.1080/17512786.2017.1397530.

Entman, R.M. (1993) 'Framing: Towards Clarification of a Fractured Paradigm', *Journal of Communication*, 43(4), pp. 51–58.

Fairclough, N. (1993) 'Critical Discourse Analysis and the Marketization of Public Discourse: The Universities', *Discourse & Society*, 4(2), pp. 133–168.

Ferguson, J. (2016) 'The secret feature of the iPhone 7 plus: experts say dual camera could be used for radical augmented reality apps', *MailOnline*, 9 September. Available at: www.dailymail.co.uk/sciencetech/article-3781859/The-secret-feature-iPhone-7-s-dual-camera-Experts-say-used-radical-augmented-reality-apps.html (Accessed: 11 February 2019).

Foxman, M. and Nieborg, D.B. (2016) 'Between a Rock and a Hard Place: Games Coverage and Its Network of Ambivalences', *Journal of Games Criticism*, 3(1).

Frewer, L.J., Scholderer, J. and Bredahl, L. (2003) 'Communicating about the Risks and Benefits of Genetically Modified Foods: The Mediating Role of Trust', *Risk Analysis*, 23(6), pp. 1117–1133. doi:10.1111/j.0272-4332.2003.00385.x.

Fuchs, P., Guez, J., Hugues, O., Jégo, J., Kemeny, A. and Mestre, D. (2017) *Virtual Reality Headsets — A Theoretical and Pragmatic Approach*. London: CRC Press.

Gibbs, S. (2014) 'Sony's Project Morpheus brings virtual reality to mainstream console gaming', *The Guardian*, 12 May. Available at: www.theguardian.com/technology/2014/may/12/sonys-project-morpheus-virtual-reality-console-gaming (Accessed: 19 December 2018).

Gitlin, T. (1980) *The Whole World Is Watching*. Berkeley: University of California Press.

Go, E., Jung, E.H. and Wu, M. (2014) 'The Effects of Source Cues on Online News Perception', *Computers in Human Behavior*, 38, pp. 358–367. doi:10.1016/j.chb.2014.05.044.

Goode, E. and Ben-Yehuda, N. (1994) 'Moral Panics: Culture, Politics, and Social Construction', *Annual Review of Sociology*, 20, pp. 149–171.

Gray, R. (2015) 'The end of air sickness? Virtual reality headsets could prevent nausea on bumpy flights and even tackle jet lag', *MailOnline*, 21 April. Available at: www.dailymail.co.uk/sciencetech/article-3048511/The-end-air-sickness-Virtual-reality-headsets-prevent-nausea-bumpy-flights-tackle-jet-lag.html   (Accessed: 13 February 2019).

Greengard, S. (2019) *Virtual Reality*. Cambridge, MA: The MIT Press.

Griffiths, S. (2014a) 'Google to take on Facebook's Oculus Rift: firm set to spend $500m on 'cinematic reality' firm Magic Leap', *MailOnline*, 14 October. Available at: www.dailymail.co.uk/sciencetech/article-2792411/google-set-spend-500m-cinematic-reality-firm-magic-leap-head-head-facebook-s-oculus-rift.html (Accessed: 14 February 2019).

Griffiths, S. (2014b) 'Is Google Glass a flop? Developers – and customers – are ditching the smart spectacles in favour of Oculus Rift', *MailOnline*, 17 November. Available at: www.dailymail.co.uk/sciencetech/article-2837750/Is-Google-Glass-flop-Developers-customers-ditching-smart-spectacles-favour-Oculus-Rift.html (Accessed: 14 February 2019).

Griffiths, S. (2016) 'PlayStation VR to launch in autumn: gaming retailer lets release date slip – but there's no confirmation about the price', *MailOnline*, 17 February. Available at: www.dailymail.co.uk/sciencetech/article-3450879/PlayStation-VR-launch-autumn-Gaming-retailer-lets-release-date-slip-s-no-confirmation-price.html (Accessed: 13 February 2019).

Griffiths, S. and Woollaston, V. (2016) 'HTC Vive price confirmed: full-body virtual reality system costs $800 – and lets users make calls from their imaginary world', *MailOnline*, 21 February. Available at: www.dailymail.co.uk/sciencetech/article-3456123/HTC-Vive-price-confirmed-body-virtual-reality-costs-800-lets-users-make-calls-imaginary-world.html (Accessed: 13 February 2019).

Hall, S., Critcher, C., Jefferson, T., Clarke, J. and Roberts, B. (1978) *Policing the Crisis*. London: Macmillan.

Hallahan, K. (1999) 'Seven Models of Framing: Implications for Public Relations', *Journal of Public Relations Research*, 11(3), pp. 205–242. doi:10.1207/s1532754xjprr1103_02.

Harcup, T. and O'Neill, D. (2017) 'What Is News? News Values Revisited (Again)', *Journalism Studies*, 18(12), pp. 1470–1488. doi:10.1080/1461670X.2016.1150193.

Hedman, J. and Gimpel, G. (2010) 'The Adoption of Hyped Technologies: A Qualitative Study', *Information Technology and Management*, 11(4), pp. 161–175. doi:10.1007/s10799-010-0075-0.

Hern, A. (2014) 'Facebook's Oculus deal is Kickstarter's first billion-dollar exit', *The Guardian*, 26 March. Available at: www.theguardian.com/technology/2014/mar/26/facebook-oculus-deal-kickstarter-first-billion-dollar-exit   (Accessed: 13 December 2018).

Hern, A. (2015) 'Will 2016 be the year virtual reality gaming takes off?', *The Guardian*, 28 December. Available at: www.theguardian.com/technology/2015/dec/28/virtual-reality-gaming-takes-off-2016 (Accessed: 21 December 2018).

Hern, A. and Gibbs, S. (2014) 'Samsung's Gear VR headset is Oculus Rift for smartphone', *The Guardian*, 3 September. Available at: www.theguardian.com/technology/2014/sep/03/samsungs-gear-vr-headset-is-oculus-rift-for-smartphone (Accessed: 19 December 2018).

Hetland, P. (2012) 'Internet between Utopia and Dystopia', *Nordicom Review*, 33(2), pp. 3–15.

Johnson, M., Egelman, S. and Bellovin, S.M. (2012) 'Facebook and Privacy: It's Complicated', *Proceedings of the Eighth Symposium on Usable Privacy and Security*, Washington, USA, 11–13 July, pp. 1–15. doi:10.1145/2335356.2335369.

Kim, H., Chan, H.C. and Gupta, S. (2007) 'Value-based Adoption of Mobile Internet: An Empirical Investigation', *Decision Support Systems*, 43(1), pp. 111–126. doi:10.1016/j.dss.2005.05.009.

Kitching, C. (2016) 'The future of in-flight entertainment? New headsets display HD films which block out annoying fellow passengers', *MailOnline*, 24 February. Available at: www.dailymail.co.uk/travel/travel_news/article-3462199/Skylights-Theater-headsets-trials-XL-Airways.html (Accessed: 13 February 2019).

Kristensen, N.N., Hellman, H. and Riegert, K. (2019) 'Cultural Mediators Seduced by *Mad Men*: How Cultural Journalists Legitimized a Quality TV Series in the Nordic Region', *Television & New Media*, 20(3), pp. 257–274. doi:10.1177/1527476417743574.

Landi, M. (2017) 'OPTICAL ILLUSION Google Glass making a surprise return after three years of silence — but would you wear one?', *Sun*, 22 June. Available at: www.thesun.co.uk/tech/3858059/google-glass-making-a-surprise-return-after-three-years-of-silence-but-would-you-wear-one/ (Accessed: 20 February 2019).

Lee, Y. and O'Connor, G.C. (2003) 'The Impact of Communication Strategy on Launching New Products: The Moderating Role of Product Innovativeness', *Journal of Product Innovation Management*, 20, pp. 4–21.

Macdonald, C. (2016) 'The English seaside town that could beat depression around the world: Researchers recreate Wembury and say a virtual visit makes patients relax', *MailOnline*, 22 January. Available at: www.dailymail.co.uk/sciencetech/article-3411028/The-English-seaside-town-beat-depression-world-Research ers-recreate-Wembury-say-virtual-visit-makes-patients-relax.html (Accessed: 13 February 2019).

Maisch, B., Binder, J., Schmid, B. and Leifer, L. (2011) 'The Dimensions of Trust — Building Confidence through Innovation Communication', *Innovation Journalism*, 8(1), pp. 1–28.

Marwick, A.E. (2008) 'To Catch a Predator? The MySpace Moral Panic', *First Monday*, 13(6). doi:10.5210/fm.v13i6.2152.

Meta (2023) *Meta Reports Fourth Quarter and Full Year 2022 Results*. 1 February [Press Release]. Available at: https://investor.fb.com/investor-news/press-release-details/2023/Meta-Reports-Fourth-Quarter-and-Full-Year-2022-Results/default.aspx (Accessed: 5 June 2023).

Milk, C. (2015) *How Virtual Reality Can Create the Ultimate Empathy Machine*. Available at: www.ted.com/talks/chris_milk_how_virtual_reality_can_create_the_u ltimate_empathy_machine (Accessed: 11 March 2020).

Morris, C. (2016) 'Sen. Al Franken takes aim at Oculus Rift privacy policies', *Fortune*, 8 April. Available at: http://fortune.com/2016/04/08/senator-al-franken-oculus-rift-privacy-policies/ (Accessed: 28 November 2016).

Narain, J. (2012) 'Surf the net, email, make calls – with your glasses! How the Google goggles work', *MailOnline*, 24 February. Available at: www.dailymail.co.uk/scie ncetech/article-2105628/Google-glasses-Surf-net-email-make-calls--Google-gogg les-work.html (Accessed: 20 February 2019).

Naughton, J. (2013) 'Are you a Google Glass half full or half empty kind of person?', *The Guardian*, 5 May. Available at: www.theguardian.com/technology/2013/may/05/google-glass-john-naughton (Accessed: 1 March 2019).

Newman, N. (2012) *Reuters Institute Digital News Report 2012*. Available at: http://media.digitalnewsreport.org/wp-content/uploads/2012/05/Reuters-Institute-Digi tal-News-Report-2012.pdf (Accessed: 7 March 2020).

Nisbet, M.C. (2010) 'Knowledge into Action: Framing the Debates Over Climate Change and Poverty', in D'Angelo, P. and Kuypers, J.A. (eds.) *Doing News Framing Analysis*. Oxon: Routledge, pp. 43–83.

Nordfors, D. (2009) 'Innovation Journalism, Attention Work and the Innovation Economy', *Innovation Journalism*, 6(1), pp. 1–46.

Pan, Z. and Kosicki, G.M. (1993) 'Framing Analysis: An Approach to News Discourse', *Political Communication*, 10, pp. 55–75.

Parkin, S. (2014) 'Comfortably numb: how virtual reality can soothe nervous patients', *The Guardian*, 7 August. Available at: www.theguardian.com/technol ogy/2014/aug/07/how-virtual-reality-can-soothe-nervous-patients (Accessed: 13 December 2018).

Parkin, S. (2016a) 'After the success of Pokémon Go!, what is the future for augmented reality?', *The Guardian*, 23 October. Available at: www.theguardian.com/technol ogy/2016/oct/23/augmented-reality-development-future-smartphone (Accessed: 13 December 2018).

Parkin, S. (2016b) 'PlayStation boss: virtual reality throws out the game-design rule book', *The Guardian*, 17 March. Available at: www.theguardian.com/tec hnology/2016/mar/16/playstation-virtual-reality-vr-headset-game-andrew-house (Accessed: 19 December 2018).

Prigg, M. (2014a) Facebook buys virtual reality headset firm Oculus for $2bn as Mark Zuckerberg promises to 'change the way we communicate', *MailOnline*, 25 March. Available at: www.dailymail.co.uk/sciencetech/article-2589367/Get-ready-social-platform-Facebook-buys-virtual-reality-firm-Oculus-2bn.html (Accessed: 14 February 2019).

Prigg, M. (2014b) 'Facebook's Oculus Rift Virtual Reality headset to cost just $200 – and could be on sale next year', *MailOnline*, 5 September. Available at: www.dailym ail.co.uk/sciencetech/article-2745488/Facebooks-Oculus-Virtual-Reality-headset-cost-just-200-sale-year.html (Accessed: 14 February 2019).

Prigg, M. (2014c) 'Is this what the Oculus Rift will look like? Designer reveals 'ski goggle' concept for Facebook's $200 VR headset', *MailOnline*, 26 December. Available at: www.dailymail.co.uk/sciencetech/article-2887963/Is-gadget-look-like-Designers-reveal-ski-goggle-concept-Facebooks-200-Oculus-Rift-VR-headset. html (Accessed: 14 February 2019).

Prigg, M. (2015a) 'Virtual Reality is coming: Facebook's Oculus Rift headset will finally ship in 2016 as firm promises it will 'transform entertainment and commu-nication'', *MailOnline*, 6 May. Available at: www.dailymail.co.uk/sciencetech/arti cle-3070674/Virtual-Reality-coming-Facebook-s-Oculus-Rift-headset-finally-ship-2016-firm-promises-transform-entertainment-communication.html (Accessed: 13 February 2019).

Prigg, M. (2015b) 'Facebook's Oculus Rift headset will cost $1500 (including the new computer you'll probably need to power it)', *MailOnline*, 27 May. Available at: www.dailymail.co.uk/sciencetech/article-3100064/Facebook-s-Oculus-Rift-headset-cost-1500-including-new-computer-ll-probably-need-power-it.html (Accessed: 13 February 2019).

Prigg, M. (2016a) 'Get ready for the Rift: Oculus to open preorders for VR headset on January 6th – but you'll probably need an expensive new PC to run it (and it could even harm your health)', *MailOnline*, 4 January. Available at: www.dailym ail.co.uk/sciencetech/article-3384239/Oculus-reveals-preorders-Rift-VR-head set-open-January-6th-ll-probably-need-expensive-new-PC-run-harm-health.html (Accessed: 13 February 2019).

Prigg, M. (2016b) 'Oculus Rift will cost $599 (and you'll probably need a new $1,000 PC as well): preorders open for Facebook's highly anticipated VR headset', *MailOnline*, 6 January. Available at: www.dailymail.co.uk/sciencetech/article-3387 222/Oculus-Rift-cost-599-ll-probably-need-new-1-000-PC-Preorders-open-highly-anticipated-VR-headset.html (Accessed: 13 February 2019).

Prigg, M. (2016c) 'Sony's PlayStation VR to undercut Oculus and HTC: headset will cost $399 and you WON'T need an expensive new PC to use it', *MailOnline*, 15 March. Available at: www.dailymail.co.uk/sciencetech/article-3494127/Sony-s-PlayStation-VR-undercut-Oculus-HTC-Headset-cost-399-WON-T-need-expens ive-new-PC-use-it.html (Accessed: 12 February 2019).

Prigg, M. (2016d) 'Mark Zuckerberg reveals Virtual Reality Facebook: Billionaire CEO chats to his wife, checks on his dog and takes a family snap with a virtual selfie stick in blockbuster demo', *MailOnline*, 6 October. Available at: www. dailymail.co.uk/sciencetech/article-3825715/Facebook-gets-virtual-Mark-Zuckerb erg-reveals-version-social-network-s-VR-software-selfie-stick.html (Accessed: 11 February 2019).

Prigg, M. (2017) 'Apple's 2019 iPhone 'to have rear facing laser 3D camera' to make augmented reality even more realistic', *MailOnline*, 15 November. Available at: www.dailymail.co.uk/sciencetech/article-5083459/Apple-s-2019-iPhone-rear-facing-laser-3D-camera.html (Accessed: 12 December 2018).

Reuters (2016) 'Searching for real growth, U.S. companies turn to virtual reality', *MailOnline*, 17 May. Available at: www.dailymail.co.uk/wires/reuters/article-3595 229/Searching-real-growth-U-S-companies-turn-virtual-reality.html (Accessed: 12 February 2019).

Rogers, E.M. (2003) *Diffusion of Innovations*. 5th edn. New York: Free Press.

Ruef, A. and Markard, J. (2010) 'What Happens after a Hype? How Changing Expectations Affected Innovation Activities in the Case of Stationary Fuel Cells', *Technology Analysis & Strategic Management*, 22(3), pp. 317–338. doi:10.1080/ 09537321003647354.

Sääksjärvi, M. and Morel, K.P.N. (2010) 'The Development of a Scale to Measure Consumer Doubt toward New Products', *European Journal of Innovation Management*, 13(3), pp. 272–293. doi:10.1108/14601061011060120.

Schäfer, M.S. (2017) 'How Changing Media Structures Are Affecting Science News Coverage', in Jamieson, K.H., Kahan, D. and Scheufele, D. (eds.) *Oxford Handbook on the Science of Science Communication*. New York: Oxford University Press, pp. 51–60.

Schatzel, K. and Calantone, R. (2006) 'Creating Market Anticipation: An Exploratory Examination of the Effect of Preannouncement Behavior on a New Product's Launch', *Journal of the Academy of Marketing Science*, 34(3), pp. 357–366. doi:10.1177/0092070304270737.

Scheufele, D.A. (2013) 'Communicating Science in Social Settings', *PNAS*, 110(3), pp. 14040–14047. doi:10.1073/pnas.1213275110.

Sedghi, A. (2014) 'Facebook: 10 Years of Social Networking, in Numbers', *The Guardian*, 4 February. Available at: www.theguardian.com/news/datablog/2014/ feb/04/facebook-in-numbers-statistics (Accessed: 9 October 2020).

Shoemaker, P.J. and Reese, S.D. (2014) *Mediating the Message in the 21st Century*. Oxon: Routledge.

Shoffman, M. (2014) 'Wearable technology like Google Glass is tipped to become a multi-billion pound industry – but is it a good fit for investors?', *MailOnline*,

21 November. Available at: www.dailymail.co.uk/money/investing/article-2828 680/Wearable-technology-like-Google-Glass-tipped-multi-billion-pound-industry-good-fit-investors.html (Accessed: 14 February 2019).

Sørensen, E. (2012) 'Violent Computer Games in the German Press', *New Media & Society*, 15(6), pp. 963–981. doi:10.1177/1461444812460976.

Steinicke, F. (2016) *Being Really Virtual*. Cham, Switzerland: Springer.

Tucker, I. (2015) 'Google Cardboard: a VR headset you make yourself', *The Guardian*, 12 June. Available at: www.theguardian.com/technology/2015/jun/12/google-cardboard-virtual-reality-vr-headset (Accessed: 20 December 2018).

van Dijk, T.A. (1988) *News as Discourse*. New Jersey: Lawrence Erlbaum Associates.

Van Gorp, B. (2007) 'The Constructionist Approach to Framing: Bringing Culture Back In', *Journal of Communication*, 57, pp. 60–78. doi:10.1111/j.1460-2466.2006.00329.x.

Van Gorp, B. (2010) 'Strategies to Take Subjectivity Out of Framing Analysis', in D'Angelo, P. and Kuypers, J.A. (eds.) *Doing News Framing Analysis*. Oxon: Routledge, pp. 84–109.

Vichiengior, T., Ackermann, C. and Palmer, A. (2019) 'Consumer Anticipation: Antecedents, Processes and Outcomes', *Journal of Marketing Management*, 35(1–2), pp. 130–159. doi:10.1080/0267257X.2019.1574435.

Vollans, E., Janes, S., Therrien, C. and Arsenault, D. (2017) 'Introduction: "It's [not Just] in the Game": the Promotional Context of Video Games', *Kinephanos*, 7(November), pp. 1–6.

Volpicelli, G. (2015) 'Apple buys Faceshift: facial recognition firm could make iPhones more secure and help develop virtual reality devices', *MailOnline*, 25 November. Available at: www.dailymail.co.uk/sciencetech/article-3333511/Apple-buys-Faceshift-Facial-recognition-firm-make-iPhones-secure-help-develop-virtual-reality-devices.html (Accessed: 13 February 2019).

Whitton, N. and Maclure, M. (2015) 'Video Game Discourses and Implications for Game-based Education', *Discourse: Studies in the Cultural Politics of Education*, 38(4), pp. 1–13. doi:10.1080/01596306.2015.1123222.

Williams, D. (2003) 'The Video Game Lightning Rod', *Information, Communication & Society*, 6(4), pp. 523–550. doi:10.1080/1369118032000163240.

Wong, J.C. (2016) 'Oculus founder apologises to VR fans over Rift price', *The Guardian*, 7 January. Available at: www.theguardian.com/technology/2016/jan/07/oculus-founder-apologises-to-vr-fans-over-rift-price (Accessed: 19 December 2018).

Woollaston, V. (2013) 'Is this the best holiday gadget ever? The Google Glass-style visor that translates ANY foreign menu and sign immediately', *MailOnline*, 30 September. Available at: www.dailymail.co.uk/sciencetech/article-2439232/NTT-Docomo-unveils-Google-Glass-style-visor-translates-foreign-menus-signs.html (Accessed: 6 December 2018).

Woollaston, V. (2014) 'Turn your mobile into a VIRTUAL REALITY HEADSET: £40 harness uses phone's sensors to track movement and play games', *MailOnline*, 21 May. Available at: www.dailymail.co.uk/sciencetech/article-2634943/Turn-mobile-VIRTUAL-REALITY-HEADSET-40-harness-uses-phones-sensors-track-movement-play-games.html (Accessed: 14 February 2019).

Woollaston, V. (2015) 'Zuckerberg believes virtual reality will become the new reality – and that we'll plug into it by wearing smart glasses', *MailOnline*, 15 May. Available at: www.dailymail.co.uk/sciencetech/article-3083057/Facebook-believes-VR-new-reality-ll-plug-wearing-smartglasses.html (Accessed: 13 February 2019).

Woollaston, V. (2016a) 'HTC reveals it sold 15,000 Vive VR headsets in the first 10 MINUTES of going on sale', *MailOnline*, 2 March. Available at: www.dailymail.co.uk/sciencetech/article-3472770/HTC-reveals-sold-15-000-Vive-VR-headsets-10-MINUTES-going-sale.html (Accessed: 13 February 2019).

Woollaston, V. (2016b) 'Google cardboard launches in the UK: basic virtual reality headset costs £15 – or you can just make your own', *MailOnline*, 11 May. Available at: www.dailymail.co.uk/sciencetech/article-3584992/Google-Cardboard-launches-UK-Basic-virtual-reality-headset-costs-15-just-make-own.html (Accessed: 12 February 2019).

Zolfagharifard, E. (2014) 'Could virtual reality prevent depression in ASTRONAUTS? Star Trek-style holodecks may help them escape the isolation of space', *MailOnline*, 15 October. Available at: www.dailymail.co.uk/sciencetech/article-2793768/could-virtual-reality-prevent-depression-astronauts-star-trek-style-holodecks-help-escape-isolation-space.html (Accessed: 14 February 2019).

8 Conclusion

As well as presenting the first extensive study of XR media discourse, this book has also provided a methodological framework that can be used by other researchers when analysing the media coverage of emerging technologies. This final chapter concludes by summarising the key findings of the study, discussing what makes these results problematic and why XR news might be this way. It then provides some more detail about the model of frame categories and how it can be applied in future research.

Key Findings

The twofold aim of the study presented in this book was to examine the news coverage of XR and the extent to which this news coverage acted as a promotional tool for XR. These aims were achieved by applying a multimodal, mixed methods approach to XR news articles and marketing materials. Informed by framing theory, this research utilised quantitative content analysis and qualitative framing analysis to examine the news and marketing of XR technologies. Four key findings stand out in the results, as will now be discussed.

Favourable Framing of XR

Overall, there was a lack of critical news coverage of XR. Positive terms were used more than negative terms in all news outlets and, while concerns and ailments were mentioned in several articles, these were very rarely the focus of any. Moreover, the frames journalists applied to XR represent the technology in a positive light. Indeed, the most used frame in all news outlets was Immersive. Immersion is the main aim and unique selling point of VR (Evans, 2019). Therefore, by framing XR as Immersive, the news articles suggest that the technology is successful in achieving its main aim, thus presenting it positively. Furthermore, the Transcendent frame involved emphasising how XR could *improve* upon what was possible with previous technology. Relatedly, the Revolutionary and Transformative frame was used to portray XR as able

DOI: 10.4324/9781003375814-8

to bring about meaningful and positive change, rather than disruption. The Different and Unique frame positively evaluated the supposed uniqueness of the technology, while the Advanced and High-Quality frame highlighted the superiority of XR. Moreover, the Much-Anticipated frame was used to generate excitement for XR products. In addition, when a positive frame had a clear opposite (e.g. comfortable versus uncomfortable), words that could counter such a frame were consistently used in smaller portions of articles than those that indicate the presence of the positive frame. Each of these points demonstrate the preference for positive news coverage of XR.

The news media are the public's main source of information about emerging technologies (Scheufele and Lewenstein, 2005; Whitton and Maclure, 2015; Williams, 2003). This means that they can have much influence on public opinion in the early stages of the diffusion process (Rogers, 2003; Scheufele and Lewenstein, 2005; Tidd, 2010). Therefore, these frames could have significant impact on how the technology is constructed in the minds of the public. Focusing on positive representations and paying little attention to the concerns, risks and social implications surrounding XR benefits the companies creating these products because it avoids critical public debate in favour of celebratory coverage.

This finding is in line with previous research on news coverage of other emerging technologies (Anderson et al., 2005; Brennen, Howard and Nielsen, 2020b; Chuan, Tsai and Cho, 2019; Cogan, 2005; Hetland, 2012; Lewenstein, Gorss and Radin, 2005; Rössler, 2001). However, it differs from news coverage of VR's main commercial application (videogames (McKernan, 2013; Whitton and Maclure, 2015)) and fictional representations of VR (Bailenson, 2018; Chan, 2014; Steinicke, 2016). Moreover, these findings show that, unlike for some other technologies (Dwyer and Stockbridge, 1999; Goggin, 2010; Lemish, 2015; Marwick, 2008), there certainly has not been a moral panic created by the media surrounding XR. Despite the range of concerns that exist regarding this technology discussed in Chapter 2, very little attention has been paid to these areas. This coincides with De Keere, Thunnissen and Kuipers (2020) analysis of binge-watching in which they found that this activity that is clearly linked to addiction was legitimised in US news rather than made the subject of a moral panic. De Keere, Thunnissen and Kuipers note that, while a moral panic was created surrounding television when it was first introduced, the same has not happened for binge-watching which appears more obviously worthy of a moral panic. Similarly, while a moral panic was created about videogames which focused on concerns of social isolation and aggression, the same has not occurred for XR, despite VR's main application being videogames and it requiring the user to block out their view of the real world with a headset to function.

It is beyond the scope of this book to hypothesise why some emerging technologies generate moral panics and some do not. However, what is significant here is that this news coverage has not only avoided creating a moral panic around XR but it has paid very little attention to critical issues surrounding XR at all. Although moral panics have been found to result in the overregulation of technologies (Marwick, 2008), the lack of critical attention paid to XR appears to have had the opposite effect. Even in 2023, although organisations such as Tech Ethos in Europe and the XR Safety Initiative in the US have been calling for XR regulation (Vinders and Howkins, 2023; XR Safety Initiative, 2021), no new policies have been developed specifically for XR technologies. That is not to say that moral panics are desirable. However, the near absence of critical coverage has reduced the perceived need for regulation, thus granting XR companies significant control over how their products are used.

Reinforcing Promotional Frames

The second major finding in this study is that many of the frames present in XR news also appeared in XR marketing. Chapters 4–6 examined these frames in detail. The following frames were shared between the two discourses: Immersive; Transcendent; Different and Unique; Revolutionary and Transformative; Advanced and High-Quality; Social; Easy to Use; and Comfortable. Moreover, the news articles even included some of the same framing devices as the marketing to construct every one of these frames. For the Immersive frame, both samples used similar imagery to depict presence. Secondly, the concept of "going beyond" was shared for the Transcendent frame. Additionally, products were described as the first of their kind within both samples to construct the Different and Unique frame. The "future of" phrase was used to frame XR as Revolutionary and Transformative and both the news and marketing positively evaluated product specifications to construct the Advanced and High-Quality frame. For the Social frame, the news and marketing referenced the concept of telepresence. Furthermore, interaction was presented as "natural" in the two discourses when employing the Easy to Use frame. Finally, both samples mentioned the effective distribution of weight to depict the Comfortable frame.

This finding is significant for three main reasons. Firstly, it indicates that the news articles have been influenced by the marketing of XR, or at least the organisations creating the marketing materials. This is supported by quantitative data showing that the creators of XR applications and devices were the most used sources within the news articles. Secondly, regardless of whether the marketing has influenced the news or not, when frames are confirmed by further information (such as appearing in two types of media) or congruent framing devices, they become harder to contest (Van Gorp, 2007), thus enhancing their persuasive power. That is to say, because the same frames and framing devices have been used in both the news and marketing, the frames themselves become more influential. Therefore, the news reinforces the frames that are present in the marketing discourse – and vice versa – making them more likely to be accepted as fact. Thirdly, since the purpose of marketing is, ultimately, to sell a product or service, these texts will clearly aim to frame XR in a way that makes it more desirable to potential consumers. In that case, since these frames are also present in the news, this effectively aids the promotion of XR. Indeed, further evidence of this is indicated by the preference for positive frames as discussed in the previous section.

This suggests that a discourse of consumerism exists in XR news, relating to Fairclough's (1993) concept of marketisation. Several other studies uncovered a blurring between news and promotional content (Chyi and Lee, 2018; Erjavec, 2004; Harro-Loit and Saks, 2006; Lewis, Williams and Franklin, 2008; Pander Maat, 2007; Sissons, 2012) and Arik and Çağlar (2005) identified discourses encouraging consumption in Turkish lifestyle news. In line with those other studies, this research found a blurring of the boundary between news and promotional content, pointing to the commercialisation of technology news. This compromises the journalistic principles of impartiality and maintaining the separation between editorial and promotion. In the current study, this is perhaps even more concerning because it has been observed not just by copying and pasting press release content, but through the use of the same frames as the marketing materials. This means that both discourses reinforce each other. In effect, the news becomes a promotional tool. Lewis, Williams and Franklin (2008) argue that neglecting to distinguish between news and promotion compromises the independence of the press. Indeed, it is important to remember that this study purposely omitted news articles that were classed as reviews, meaning that readers would expect they are accessing news that presents facts rather than opinions (Pan and Kosicki, 1993). Therefore, this is a concerning result regarding the integrity of news coverage about emerging technologies. Such coverage, while misleading to readers, benefits XR companies by increasing the reach of their promotional frames in a context that disguises them as factual news.

XR Companies as Frame Advocates

The third main finding of this study is that the creators of XR devices and applications played a major role as advocates in the frame-building process. Content analysis revealed that application creators and device creators were used as sources in much larger portions of articles than any other source type. Additionally, the largest portion of multimedia were attributed to device creators. While multimedia attributed to news agencies were the second most common, application creators were the third most used. Therefore, it is clear that the news outlets have allowed these source types to be the primary definers (Critcher, 2003; Hall et al., 1978) of XR, both through the written word and visually. These two groups are invested in the success of XR and are therefore unlikely to be critical of the technology. Instead, they would be advocates of frames that represent XR positively. Being a news source allows social actors access to persuasive influence and gives them the power to define reality (Carlson and Franklin, 2011; Coleman and Ross, 2010).

Indeed, the prevalence of positive frames, plus the shared frames between XR news and marketing (some of which is produced by these groups) suggests they have been successful in getting their favoured definitions of XR to dominate the news coverage.

In particular, one individual was instrumental in the framing of XR: Mark Zuckerberg. Zuckerberg appears to have been a driving force in the attention paid to XR by the *Guardian* and *MailOnline*, since these news outlets first started reporting substantially on the topic in 2014 – the year Facebook/ Meta purchased Oculus. Indeed, the Oculus Rift VR headset was mentioned in, by far, the most articles in comparison with other devices. Moreover, statements from Zuckerberg were used as framing devices to construct five different frames: Revolutionary and Transformative; Advanced and High-Quality; Social; Easy to Use; and Important. These quotes and citations were also usually repeated in multiple articles, which increased the strength of those frames. Such sourcing practices give Zuckerberg power. As Carlson states, "[f]or a news story to include an individual or an organization as a source is not a neutral act but one that bestows authority through granting the source the right to be listened to" (2017: 132). Thus, Zuckerberg has repeatedly been given the authority to define XR by these news outlets, highlighting his power as a frame advocate. Whereas journalists in the fourth estate role should hold those in power to account (McNair, 2009), these sourcing practices afford even greater power to elites (in this case, technology company owners – Zuckerberg in particular), benefitting them more so than the general public.

News Promotes the Diffusion of XR

Much evidence was uncovered to suggest that the news promotes the diffusion of XR. Based on diffusion of innovations theory and models of technological acceptance, the majority of frames used in XR news positively emphasise an aspect of an innovation or new technology that makes it more likely to be adopted. In more detail, both the Transcendent and Advanced and High-Quality frames highlight the relative advantage (Rogers, 2003) of XR. The Easy to Use frame positively emphasises the ease of use (Buenaflor and Kim, 2013; Davis, 1989), complexity (Rogers, 2003) and technicality (Kim, Chan and Gupta, 2007) of XR. The Social frame supports the perceived compatibility (Rogers, 2003) of XR, while the Affordable frame assures that the perceived fee (Kim, Chan and Gupta, 2007) of XR is acceptable. Moreover, the Comfortable frame positively evaluates the physical comfort (Buenaflor and Kim, 2013) of the devices. The uniqueness of an innovation has been shown to be another factor enhancing its adoption rate (Cooper, 1979; Flight et al., 2011), meaning the Different and Unique frame could also contribute to supporting the diffusion of XR. In addition to specific frames, the prominence of entertainment applications emphasises the enjoyment benefit (Kim, Chan and Gupta, 2007) of the technology. Likewise, the focus on applications and devices as article topics improves the observability (Rogers, 2003) of XR. The lack of coverage about risks or concerns could also support diffusion because technologies that are perceived as posing risks are less likely to be adopted (Buenaflor and Kim, 2013).

Furthermore, while not a specific characteristic, Rogers (2003) argues that the higher the perceived importance of an innovation, the more likely it is to be adopted. The Revolutionary and Transformative and Important frames both emphasise this importance, thus promoting adoption. Similarly, the Successful frame helps to reduce the uncertainty about XR, which is a major part of the innovation-decision process (Rogers, 2003). Finally, the Much-Anticipated frame works to raise expectations about XR, which can support its adoption (Hedman and Gimpel, 2010), though this might lead to disappointment later on (Ruef and Markard, 2010). The Immersive frame is the only one that does not obviously link to diffusion or technological acceptance theories. However, as mentioned above, immersion is the main selling point of VR (Evans, 2019). Therefore, emphasising this could indeed support its adoption.

Added to this, Chapter 3 showed that some news articles included information about how or where to purchase XR products, even in the form of links to retailers. Such practices directly support the diffusion of the technology. Moreover, this indicates that these news outlets may have some financial incentive for framing XR so positively. Indeed, the *Guardian* even noted that it could earn commission if the reader made a purchase after clicking on such a link. Although the *Guardian* claims that this does not compromise their journalistic independence, the favourable frames suggest otherwise. If these news outlets gain money when their readers purchase XR products, they would be more likely to frame the technology positively so as to encourage these purchases. This would explain the overall promotional tone of the news articles. These results align with Chyi and Lee's (2018) study of tablets and smartphones, in which they argue that technology news is commercialised. It appears that, when it comes to news coverage of XR, the commercial agendas of the news outlets have caused them to frame the technology in a way that aligns with the interests of the industry rather than the general public.

From "Better Than Life" to News as a Promotional Tool

As stated in Chapter 1, my enquiry into news coverage of XR was initially spurred by an article encouraging escapism into virtual worlds by representing VR experiences as superior to reality. While this type of discourse was somewhat present in the news articles through the use of the Transcendent frame, presenting XR as better than real life was not a common trope. On the other hand, this study uncovered another (related) concern in XR news coverage: these technologies are presented positively, in line with the way they are marketed, leading to the news acting as a promotional tool for these products. Although this is a different concern to the one I started out with, the lack of critical coverage still encourages escapism into these virtual worlds, even if it is not by claiming the experience is superior to being in the real world. Furthermore, this highlights that technology news, at least surrounding XR, does not maintain the journalistic independence required for its fourth estate role (Hampton, 2010). Instead, XR news has more in common with other genres of journalism, such as lifestyle journalism, which has been found to include messages that encourage consumption (Arik and Çağlar, 2005) and is seen by some as an extension of marketing (English and Fleischman, 2019; Kristensen, Hellman and Riegert, 2019).

News coverage of emerging technologies can shape public debate which, in turn, affects regulation and policy decisions (Marwick, 2008; Schäfer, 2017; Scheufele, 2013). However, this promotional XR coverage encourages audiences to adopt these technologies rather than consider the ethical and political concerns that surround XR. Therefore, this news does not prioritise the public as journalists should in the fourth estate model (Fjæstad, 2007), but instead benefits the large technology companies selling these products. Rather than giving power to the public by holding elite organisations to account (e.g. by challenging positive views of XR and highlighting potential concerns), the news media give power to those elite organisations by allowing their voices to dominate the news and presenting the technology in a way that aligns with their promotional framing of the products.

Media outlets must make money to continue operating, meaning news content can be affected by their organisation's commercial interests. Journalists are under increasing pressure to produce news content quickly and regularly, particularly for online platforms (Currah, 2009; Forde and Johnston, 2013; Lewis et al., 2008), sometimes resulting in a practice of "churnalism" (Davies, 2009). In the current study, such a practice was evident, particularly in the *MailOnline* which published news wire copy verbatim and copied and pasted parts of its articles from one to another. The *Guardian* also repeated the same quotes from Zuckerberg multiple times. Since the *Sun*, *Guardian* and *MailOnline* are each subject to these same commercial pressures, this could explain the lack of variation between the news outlets in the way they frame XR. With the aim to create news content quickly and before their competitors, the resulting news coverage is uncritical and lacking in diversity, giving XR companies the power to define the technology in a way that benefits them.

Another commercial factor that can impact content in online news particularly is how much attention journalists expect to receive for certain types of stories. In their study of UK technology journalists, Brennen, Howard and Nielsen (2020a) found that traffic metrics were a key factor that influenced news content. They state that journalists "seemed to have an intuitive sense that *uncritical* stories of new tech products from well-known popular brands are reliable draws" of traffic (2020a: 12–13, original emphasis). More traffic means more readers, which translates into greater revenue from advertising and subscription models. Therefore, favourable frames support the commercial interest of the news outlets, which could explain the lack of critical coverage about XR in these online outlets, as well as the focus on XR devices created by large companies such as Facebook/Meta and Google.

Aside from being large companies that might attract traffic, online news outlets have relationships with Facebook/Meta and Google, which could explain why the news coverage is this way. Most traffic to online news sites comes from search engines and social media (Currah, 2009; Newman et al., 2017) and the two companies with the biggest market shares in these areas respectively are Google and Facebook (Media Reform Coalition, 2019; Newman, 2012). As Watson (2016) states, if publications rely on such companies to reach audiences, it is unlikely they will be critical about them or their products. Such influences appear to have played a role here since the three publications in this study each mentioned the Facebook-owned Oculus Rift device most often and Zuckerberg played a major role as a frame advocate in the articles.

Furthermore, the discussion presented in this book suggests that the news outlets certainly do have something to gain by presenting XR positively and promoting diffusion. Native advertising was found to be present within some articles in the form of links to XR retailers. These news outlets would benefit most from this native advertising if readers click the link and purchase a product. Thus, it is in their commercial interest to present this technology positively to encourage adoption. Additionally, XR companies were most frequently used as sources and a handful of articles in the *Guardian* and *MailOnline* were even written by the creators of XR applications. This indicates that relationships exist between these news outlets and these groups. Within both lifestyle and games journalism, industry officials have been known to pull advertising or stop providing a news outlet with free gifts (such as technological devices) and information if the content is unfavourable to their products (Carlson, 2009; Hanusch, Hanitzsch and Lauerer, 2017). Therefore, the news outlets in this study may have avoided critical portrayals of XR in order to maintain these relationships and their commercial benefits.

Overall, it appears that the capitalist social system news organisations operate within has led them to prioritise their own commercial interests rather than the interests of the general public. As an effect, the news also supports the agendas of XR companies trying to sell this technology to consumers. Readers are treated as commodities for the commercial gain of the newsrooms, compromising the fourth estate role of journalism to provide independent information to the public and to hold those in power to account. More regulation, or at least policing of existing regulation, is needed to ensure the boundary between news and promotional content remains separate in the interests of the general public.

Recent Developments and Future Research

Since completing this research, there have been considerable developments in the XR industry, as well as for some of the key figures in the area. Sony released its second version of the PlayStation VR headset, PlayStation VR2, in February 2023 (Tomatis, 2022) and Meta's third iteration of its Quest headset, Meta Quest 3, is due for release in autumn 2023 (Meta, 2023). Most recently, in June 2023, Apple announced its first XR headset, Vision Pro, calling it a "spatial computer" that is controlled by the user's eye movements, hand gestures and voice (Apple, 2023). Rumours of Apple producing such a device were present even in the news articles in the sample of this study (e.g. Liberatore, 2017; Murphy, 2017; Prigg, 2016), perhaps indicating that the news media started to build anticipation for this product several years ago.

Moreover, one of the most significant developments in the XR industry has been Mark Zuckerberg's rebranding of Facebook to Meta and the introduction of his vision of the metaverse in 2021. In Zuckerberg's vision, the metaverse can be described as a hyper-realistic digital world that blends virtual, augmented and mixed reality to allow users to go about their daily activities in this alternate environment, with more opportunities than are possible in the physical world (see Meta, 2021). Although the metaverse has existed in different forms for many years, including the online virtual worlds of massively multiplayer online games (such as Second Life and World of Warcraft; see Bolter, Engberg and MacIntyre, 2021), the term has recently been popularised as a result of Zuckerberg's vision. Zuckerberg's concept of the metaverse will not be possible for many years due to current technological capabilities; however, several VR applications have now been launched and classified as examples of the metaverse. This includes Decentraland, The Sandbox and Meta's Horizon Worlds, which see players building their own virtual worlds online, watching live virtual concerts, socialising and many other experiences.

This concept of the metaverse appears to have brought some more negativity into the discourse surrounding XR, at least from an anecdotal perspective. For instance, a *BBC* news item detailed the abuse and sexual harassment faced by one of their reporters when they posed as a 13-year-old in the VRchat application, raising concerns over child safety (*BBC*, 2022). A *Channel 4* documentary titled *Inside the Metaverse Are You Safe? Dispatches* (2022) discussed similar issues in more detail. In this case, it might seem that the features of a moral panic that involve highlighting the safety of children (e.g. Marwick, 2008; Petley, 1984) could be emerging in relation to the metaverse, while they were absent from XR coverage in the earlier years. Still, it is not possible to definitively make this claim without further empirical research. Taking all of these considerations into account, it is possible that the news framing of XR has changed since the sample period of this study. However, the results presented in this book are valuable to show how the technologies were presented and defined during their inception. Moving forward, future studies may want to analyse more recent coverage to see how (or if) the discourse has changed.

In addition to studies of news discourse, future research could build upon the textual analysis presented in this book by utilising ethnographic approaches. While this study used diffusion theories to examine whether news coverage promoted the adoption of XR, future research could assess this by examining XR news coverage alongside framing effects (i.e. whether these frames made individuals more or less likely to purchase an XR product). In a different way, other researchers could analyse the news production process more closely by carrying out an ethnographic study of journalists as they create XR news. This would provide more accurate data as to the factors influencing the frame-building process when it comes to news about XR.

Extending beyond XR, it is also important to examine other emerging technologies, both in terms of their media coverage and the relationship between news and promotional content. This would allow comparisons to be made. Other notable emerging technologies that could be analysed include big data, non-fungible tokens (NFTs) and artificial intelligence (AI). In particular, it would be beneficial to examine the media coverage of the generative AI platforms which have suddenly gained momentous traction since late 2022 when Open AI's ChatGPT was made publicly available (Hu, 2023). The model of frame categories presented in this book would be an effective tool to use for such research. With this in mind, the model will now be discussed in more detail.

A Model of Frame Categories for Researching Emerging Technologies

Chapters 4–7 introduced the four categories of a model to examine media coverage of emerging technologies. Tying this together, Figure 8.1 depicts each frame category with examples of questions that researchers might ask during analysis to identify frames in those categories.

To add further clarification, it is likely that there will be some overlap between the specific frames that are identified for different technologies in categories 3–4. The frames that appear in the first category (Conceptualisation) are more likely to differ since it specifically considers which concepts stand out in relation to that technology in particular. For XR, it was immersion and transcendence. For other technologies such as artificial intelligence, it might be the concept of smart or intelligent computing. For a study analysing the historical emergence of smartphones, it could be the mobile nature of the devices, or the concept of convergence. As an actor in the social construction of technology, the news media play an important role in how an innovation is conceptualised (McKernan, 2013) highlighting the value of studying these types of frames.

Moreover, an important note should be made about the fourth frame category (Evaluation). Although some of the frames included in other categories could involve evaluating the technology in some sense (e.g. how comfortable a device is), this fourth frame category is broader. It allows a researcher to analyse the positive and negative portrayals of the technology in more depth without being restricted to the specifics of the other categories. The focus is on uncovering details about the main positive and/or negative representations

Conceptualisation	Newness	User Experience	Evaluation
Frames related to concepts specific to the technology under study.	*Frames highlighting what makes a technology new or different.*	*Frames related to the actual use of a technology.*	*Frames emphasising either a positive or negative aspect of a technology.*
• What specific characteristics of the technology have been emphasised within the discourse?	• What is said to be significantly new/different about this technology compared to others?	• How is the hardware and/or software presented?	• Overall, how positive/negative is the coverage?
• What are portrayed as the key features of the technology?	• What new experiences/opportunities is the technology said to afford?	• What uses of the technology are given the most/least attention?	• What is said to be good/bad about the technology?
• What concepts in the academic literature are associated with the technology and does this align with its media representations?	• What impact or changes is the technology said to bring?	• What aspects of the user experience are said to be strong/weak?	• What benefits are mentioned and to what extent are they emphasised?
	• How advanced (or not) is the technology said to be?		• What concerns are mentioned and to what extent are they emphasised?
			• How important/significant is the technology deemed to be?

Figure 8.1 A model of frame categories for identifying frames within media discourse of emerging technologies.

that appear. As the perceptions of innovations can impact their success (Buenaflor and Kim, 2013), analysing how they have been evaluated in the news is key because this can shed light on the role the media plays in creating these perceptions and potentially impacting adoption.

As noted previously, inductive approaches to framing analysis have been criticised for lacking replicability and comparability (e.g. Tankard, 2001). However, the utilisation of these frame categories allows researchers to avoid both the prescriptive nature of deductive frame identification and the issues around comparability. That is to say, it is possible to identify new and emerging frames while still being able to make comparisons between the set categories. Still, it should also be highlighted that those carrying out future studies must not be restricted by this model if frames emerge that do not fit into these categories. It is recommended that the model be adapted or developed based on other technologies and future developments. Certainly, this model has been developed based on one group of technologies at a certain point in time. Due to the changeable nature of technological developments and innovation, it is unlikely that this model will suit every future emerging technology. For now, the model provides a strong starting point to apply when researching media coverage of emerging technologies, allowing for both the freedom to identify frames inductively and to make comparisons between studies using the same framework.

Final Remarks

This book has shown that news coverage of XR is primarily positive and there are several frames shared between the news and marketing of XR. Thus, the two texts work to reinforce each other and the frames within them. This leads to an overall discourse of consumerism in XR news, which points to the commercialisation of this news. These results are similar to previous studies on other emerging technologies and investigations into the diminishing boundary between news and promotional content. However, it appears that news coverage of XR differs from its fictional representations as well as news portrayals of videogames. The study presented here has made an original contribution to the literature regarding news coverage of emerging technologies by focusing on a previously unexplored topic (XR). It also makes an original contribution to studies looking at the relationship between news and promotional content, extending such research by analysing the interplay between news and marketing in general rather than simply native advertising or public relations material. Furthermore, the book has made a theoretical and methodological contribution in the form of frames and frame categories that can be applied to future research on XR and emerging technologies. In all, this book has presented the first in-depth investigation into XR news, as well as its connection to promotional materials, through the rigorous application of a mixed methods methodology.

The results presented here show that XR news prioritises commercial interests, both of their own media organisations as well as XR companies, rather than serving the general public. This highlights a problem with technology news because it compromises the fourth estate role of journalism. More regulation is needed to ensure the boundary between editorial and promotional content is maintained. As it stands, news about XR has been affected by the capitalist ideologies of media organisations to the extent that it acts as a promotional tool for XR companies, encouraging readers to escape into virtual worlds. Through the application of the model of frame categories, future research will be able to identify whether this trend persists in more recent XR coverage and, indeed, the media coverage of other emerging technologies.

References

Anderson, A., Allan, S., Petersen, A. and Wilkinson, C. (2005) 'The Framing of Nanotechnologies in the British Newspaper Press', *Science Communication*, 27(2), pp. 200–220. doi:10.1177/1075547005281472.

Apple (2023) *Introducing Apple Vision Pro*, 5 June. Available at: https://youtu.be/TX9qSaGXFyg (Accessed: 8 June 2023).

Arik, M.B. and Çağlar, S. (2005) 'The Face of Consumption Society in the Press: Life Style Journalism', *Proceedings of the 3rd International Symposium Communication in the Millennium*, Chapel Hill, NC, 11–13 May. Available at: https://citeseerx.ist.psu.edu/viewdoc/download?doi=10.1.1.507.9260&rep=rep1&type=pdf (Accessed: 7 April 2021).

Bailenson, J. (2018) *Experience on Demand*. New York: W.W. Norton & Company.

BBC (2022) 'Undercover journalist witnesses abuse in metaverse', *BBC*, 23 February. Available at: www.bbc.co.uk/news/av/uk-60466557 (Accessed: 8 June 2023).

Bolter, J.D., Engberg, M. and MacIntyre, B. (2021) *Reality Media: Augmented and Virtual Reality*. London: MIT Press.

Brennen, S.J., Howard, P.N. and Nielsen, R.K. (2020a) 'Balancing Product Reviews, Traffic Targets, and Industry Criticism: UK Technology Journalism in Practice', *Journalism Practice*, [Preprint], pp. 1–18. doi:10.1080/17512786.2020.1783567.

Brennen, S.J., Howard, P.N. and Nielsen, R.K. (2020b) 'What to Expect When You're Expecting Robots: Futures, Expectations, and Pseudoartificial General Intelligence in UK News', *Journalism*, 5 August. doi:10.1177/1464884920947535.

Buenaflor, C. and Kim, H. (2013) 'Six Human Factors to Acceptability of Wearable Computers', *International Journal of Multimedia and Ubiquitous Engineering*, 8(3), pp. 103–114.

Carlson, M. (2017) *Journalistic Authority: Legitimating News in the Digital Era*. New York: Columbia University Press.

Carlson, M. and Franklin, B. (2011) 'Introduction', in Franklin, B. and Carlson, M. (eds.) *Journalists, Sources, and Credibility*. Oxon: Routledge, pp. 1–15.

Carlson, R. (2009) '*Too Human* Versus the Enthusiast Press: Video Game Journalists as Mediators of Commodity Value', *Transformative Works and Cultures*, 2. doi:10.3983/twc.2009.0098.

Chan, M. (2014) *Virtual Reality: Representations in Contemporary Media*. London: Bloomsbury.

Chuan, C., Tsai, W.S. and Cho, S.Y. (2019) 'Framing Artificial Intelligence in American Newspapers', *Proceedings of the 2019 AAAI/ACM Conference on AI, Ethics, and Society*, Honolulu, USA, 27–28 January, pp. 339–344. doi:10.1145/3306618.3314285.

Chyi, H.I. and Lee, A.M. (2018) 'Commercialization of Technology News', *Journalism Practice*, 12(5), pp. 585–604. doi:10.1080/17512786.2017.1333447.

Cogan, B. (2005) '"Framing Usefulness:" An Examination of Journalistic Coverage of the Personal Computer from 1982–1984', *Southern Journal of Communication*, 70(3), pp. 248–295. doi:10.1080/10417940509373330.

Coleman, S. and Ross, K. (2010) *The Media and the Public*. West Sussex: Wiley-Blackwell.

Cooper, R.G. (1979) 'The Dimensions of Industrial New Product Success and Failure', *Journal of Marketing*, 43(3), pp. 93–103. doi:10.2307/1250151.

Critcher, C. (2003) *Moral Panics and the Media*. Buckingham: Open University Press.

Currah, A. (2009) *What's Happening to Our News*. Oxford: Reuters Institute for the Study of Journalism.

Davies, N. (2009) *Flat Earth News*. London: Vintage.

Davis, F.D. (1989) 'Perceived Usefulness, Perceived Ease of Use, and User Acceptance of Information Technology', *MIS Quarterly*, 13(3), pp. 319–340.

De Keere, K., Thunnissen, E. and Kuipers, G. (2020) 'Defusing Moral Panic: Legitimizing Binge-Watching as Manageable, High-Quality, Middle-Class Hedonism', *Media, Culture & Society*, [Preprint], pp. 1–19. doi:10.1177/0163443720972315.

Dwyer, T. and Stockbridge, S. (1999) 'Putting Violence to Work in New Media Policies', *New Media & Society*, 1(2), pp. 227–249.

English, P. and Fleischman, D. (2019) 'Food for thought in Restaurant Reviews', *Journalism Practice*, 13(1), pp. 90–104. doi:10.1080/17512786.2017.1397530.

Erjavec, K. (2004) 'Beyond Advertising and Journalism: Hybrid Promotional News Discourse', *Discourse & Society*, 15(5), pp. 553–578.

Evans, L. (2019) *The Re-Emergence of Virtual Reality*. Oxon: Routledge.

Fairclough, N. (1993) 'Critical Discourse Analysis and the Marketization of Public Discourse: The Universities', *Discourse & Society*, 4(2), pp. 133–168.

Fjæstad, B. (2007) 'Why Journalists Report Science as They Do', in Bauer, M.W. and Bucci, M. (eds.) *Journalism, Science and Society: Science Communication Between News and Public Relations*. Oxon: Routledge, pp. 123–131.

Flight, R.L., Allaway, A.W., Kim, W. and D'Souza, G. (2011) 'A Study of Perceived Innovation Characteristics across Cultures and Stages of Diffusion', *Journal of Marketing Theory and Practice*, 19(1), pp. 109–126. doi:10.2753/MTP1069-6679190107.

Forde, S. and Johnston, J. (2013) 'The News Triumvirate: Public Relations, Wire Agencies and Online Copy', *Journalism Studies*, 14(1), pp. 113–129. doi:10.1080/1461670X.2012.679859.

Goggin, G. (2010) 'Official and Unofficial Mobile Media in Australia: Youth, Panics, Innovation', in Donald, S.H., Anderson, T.D. and Spry, D. (eds.) *Youth, Society and Mobile Media in Asia*. Oxon: Routledge, pp. 120–134.

Hall, S., Critcher, C., Jefferson, T., Clarke, J. and Roberts, B. (1978) *Policing the Crisis*. London: Macmillan.

Hampton, M. (2010) 'The Fourth Estate Ideal in Journalism History', in Allan, S. (ed.) *The Routledge Companion to News and Journalism*. Rev. edn. Oxon: Routledge, pp. 3–12.

Hanusch, F., Hanitzsch, T. and Lauerer, C. (2017) "How Much Love Are You Going to Give This Brand?' Lifestyle Journalists on Commercial Influences in Their Work', *Journalism*, 18(2), pp. 141–158. doi:10.1177/1464884915608818.

Harro-Loit, H. and Saks, K. (2006) 'The Diminishing Border between Advertising and Journalism in Estonia', *Journalism Studies*, 7(2), pp. 312–322. doi:10.1080/14616700500533635.

Hedman, J. and Gimpel, G. (2010) 'The Adoption of Hyped Technologies: A Qualitative Study', *Information Technology and Management*, 11(4), pp. 161–175. doi:10.1007/s10799-010-0075-0.

Hetland, P. (2012) 'Internet between Utopia and Dystopia', *Nordicom Review*, 33(2), pp. 3–15.

Hu, K. (2023) *ChatGPT Sets Record for Fastest-Growing User Base – Analyst Note*. 2 February [Press Release]. Available at: www.reuters.com/technology/chatgpt-sets-record-fastest-growing-user-base-analyst-note-2023-02-01/ (Accessed: 25 July 2023).

Inside the Metaverse Are You Safe? Dispatches (2022) Channel 4, 25 April, 20:00.

Kim, H., Chan, H.C. and Gupta, S. (2007) 'Value-based Adoption of Mobile Internet: An Empirical Investigation', *Decision Support Systems*, 43(1), pp. 111–126. doi:10.1016/j.dss.2005.05.009.

Kristensen, N.N., Hellman, H. and Riegert, K. (2019) 'Cultural Mediators Seduced by *Mad Men*: How Cultural Journalists Legitimized a Quality TV Series in the Nordic Region', *Television & New Media*, 20(3), pp. 257–274. doi:10.1177/1527476417743574.

Lemish, D. (2015) 'Media Moral Panic about Screens and Teens', *Gateway Journalism Review*, 45(339).

Lewenstein, B.V., Gorss, J. and Radin, J. (2005) 'The Salience of Small: Nanotechnology Coverage in the American Press, 1986–2004', *The Annual International Communication Association Conference*. Sheraton New York Times Square Hotel, New York, 26–30 May. Available at: https://ecommons.cornell.edu/bitstream/handle/1813/14275/LewensteinGorssRadin.2005.NanoMedia.ICA.pdf (Accessed: 10 October 2016).

Lewis, J., Williams, A. and Franklin, B. (2008) 'A Compromised Fourth Estate?', *Journalism Studies*, 9(1), pp. 1–20. doi:10.1080/14616700701767974.

Lewis, J., Williams, A., Franklin, B., Thomas, J. and Mosdell, N. (2008) *The Quality and Independence of British Journalism*. Available at: https://orca.cf.ac.uk/18439/1/Quality%20%26%20Independence%20of%20British%20Journalism.pdf (Accessed: 16 March 2018).

Liberatore, S. (2017) 'Apple hires NASA's augmented reality boss as secretive smart glasses project gets ready for blast off', *MailOnline*, 25 April. Available at: www.dailymail.co.uk/sciencetech/article-4444568/Apple-hires-NASA-s-AR-head-smart-glasses-project.html (Accessed: 8 February 2019).

Marwick, A.E. (2008) 'To Catch a Predator? The MySpace Moral Panic', *First Monday*, 13(6). doi:10.5210/fm.v13i6.2152.

McKernan, B. (2013) 'The Morality of Play: Video Game Coverage in *The New York Times* From 1980 to 2010', *Games and Culture*, 8(5), pp. 307–329. doi:10.1177/1555412013493133.

McNair, B. (2009) 'Journalism and Democracy', in Wahl-Jorgensen, K. and Hanitzsch, T. (eds.) *The Handbook of Journalism Studies*. Oxon: Routledge, pp. 237–249.

Media Reform Coalition (2019) *Who Owns the UK Media?* Available at: www.mediareform.org.uk/wp-content/uploads/2019/03/FINALonline2.pdf (Accessed: 6 August 2020).

Meta (2021) *The Metaverse and How We'll Build It Together – Connect 2021*, 28 October. Available at: https://youtu.be/Uvufun6xer8 (Accessed: 8 June 2023).

Meta (2023) *The Powerful Meta Quest 3. Coming This Autumn*. Available at: www.meta.com/gb/quest/quest-3 (Accessed: 8 June 2023).

Murphy, M. (2017) 'I GLASSES? Apple leaks suggest tech giant is working on augmented reality glasses dubbed 'Project Mirrorshades'', *Sun*, 5 June. Available at: www.thesun.co.uk/tech/3729577/apple-leaks-suggest-tech-giant-is-working-on-augmented-reality-glasses-dubbed-project-mirrorshades/ (Accessed: 7 March 2019).

Newman, N. (2012) *Reuters Institute Digital News Report 2012*. Available at: http://media.digitalnewsreport.org/wp-content/uploads/2012/05/Reuters-Institute-Digital-News-Report-2012.pdf (Accessed: 7 March 2020).

Newman, N., Fletcher, R., Kalogeropoulos, A., Levy, D.A.L. and Nielsen, R.K. (2017) *Reuters Institute Digital News Report 2017*. Available at: https://reutersinstitute.politics.ox.ac.uk/sites/default/files/Digital%20News%20Report%202017%20web_0.pdf (Accessed: 7 March 2020).

Pan, Z. and Kosicki, G.M. (1993) 'Framing Analysis: An Approach to News Discourse', *Political Communication*, 10, pp. 55–75.

Pander Maat, H. (2007) 'How Promotional Language in Press Releases Is Dealt with by Journalists', *Journal of Business Communication*, 44(1), pp. 59–95. doi:10.1177/0021943606295780.

Petley, J. (1984) 'A Nasty Story', *Screen*, 25(2), pp. 68–75. doi:10.1093/screen/25.2.68.

Prigg, M. (2016) 'Apple has 'team of hundreds' developing virtual and augmented reality headsets', *MailOnline*, 29 January. Available at: www.dailymail.co.uk/sciencetech/article-3423326/Apple-team-hundreds-developing-virtual-augmented-reality-system.html (Accessed: 13 February 2019).

Rogers, E.M. (2003) *Diffusion of Innovations*. 5th edn. New York: Free Press.

Rössler, P. (2001) 'Between Online Heaven and Cyberhell', *New Media & Society*, 3(1), pp. 49–66.

Ruef, A. and Markard, J. (2010) 'What Happens after a Hype? How Changing Expectations Affected Innovation Activities in the Case of Stationary Fuel Cells', *Technology Analysis & Strategic Management*, 22(3), pp. 317–338. doi:10.1080/09537321003647354.

Schäfer, M.S. (2017) 'How Changing Media Structures Are Affecting Science News Coverage', in Jamieson, K.H., Kahan, D. and Scheufele, D. (eds.) *Oxford Handbook on the Science of Science Communication*. New York: Oxford University Press, pp. 51–60.

Scheufele, D.A. (2013) 'Communicating Science in Social Settings', *PNAS*, 110(3), pp. 14040–14047. doi:10.1073/pnas.1213275110.

Scheufele, D.A. and Lewenstein, B.V. (2005) 'The Public and Nanotechnology: How Citizens Make Sense of Emerging Technologies', *Journal of Nanoparticle Research*, 7, pp. 659–667. doi:10.1007/s11051-005-7526-2.

Sissons, H. (2012) 'Journalism and Public Relations: A Tale of Two Discourses', *Discourse & Communication*, 6(3), 273–294. doi:10.1177/1750481312452202.

Steinicke, F. (2016) *Being Really Virtual.* Cham, Switzerland: Springer.

Tankard, J.W. (2001) 'The Empirical Approach to the Study of Media Framing', in Reese, S.D., Gandy, O.H. and Grant, A.E. (eds.) *Framing Public Life.* Oxon: Routledge, pp. 95–105.

Tidd, J. (2010) 'From Models to the Management of Diffusion', in Tidd, J. (ed.) *Gaining Momentum: Managing the Diffusion of Innovations.* London: Imperial College Press, pp. 3–45.

Tomatis, I. (2022) 'PlayStation VR2 launches in February at $549.99', *PlayStation. Blog,* 2 November. Available at: https://blog.playstation.com/2022/11/02/playstation-vr2-launches-in-february-at-549-99/ (Accessed: 8 June 2023).

Van Gorp, B. (2007) 'The Constructionist Approach to Framing: Bringing Culture Back In', *Journal of Communication,* 57, pp. 60–78. doi:10.1111/j.1460-2466.2006.00329.x.

Vinders, J. and Howkins, B. (2023) *Enhancing EU Legal Frameworks for Digital Extended Reality.* Available at: www.techethos.eu/wp-content/uploads/2023/03/TECHETHOS-POLICY-BRIEF_Digital-Extended-Reality_for-web.pdf (Accessed: 8 June 2023).

Watson, S.M. (2016) *Toward a Constructive Technology Criticism.* Tow Centre for Digital Journalism. Available at: www.cjr.org/tow_center_reports/constructive_technology_criticism.php (Accessed: 6 April 2021).

Whitton, N. and Maclure, M. (2015) 'Video Game Discourses and Implications for Game-based Education', *Discourse: Studies in the Cultural Politics of Education,* pp. 1–13. doi:10.1080/01596306.2015.1123222.

Williams, D. (2003) 'The Video Game Lightning Rod', *Information, Communication & Society,* 6(4), pp. 523–550. doi:10.1080/1369118032000163240.

XR Safety Initiative (2021) *An Imperative: Developing Standards for Safety and Security in XR Environments.* Available at: https://xrsi.org/wp-content/uploads/2021/02/An-Imperative-Emteq-XRSI-Whitepaper-on-Standards-for-safety-and-Security.pdf (Accessed: 5 May 2021).

Appendices

Appendix 1
Main Topic of Articles

Table A1.1 Main topic of articles per news outlet

Topic	Sun		Guardian		MailOnline		Overall	
	No.	*Percent*	*No.*	*Percent*	*No.*	*Percent*	*No.*	*Percent*
Application(s)	35	57.38	149	60.08	298	44.61	482	49.33
- XR focus	33	54.10	144	58.06	273	40.87	450	46.06
- With XR element	2	3.28	5	2.02	25	3.74	32	3.28
product(s)	10	16.39	32	12.90	200	29.94	242	24.77
- Commercial product(s)	5	8.20	29	11.69	135	20.21	169	17.30
- Rumoured product(s)	3	4.92	1	0.40	40	5.99	44	4.50
- Industry product(s)	2	3.28	2	0.81	20	2.99	24	2.46
- Conceptual product(s)	0	0.00	0	0.00	5	0.75	5	0.51
Demo	6	9.84	4	1.61	47	7.04	57	5.83
- General public	2	3.28	1	0.40	35	5.24	38	3.89
- Journalist	1	1.64	3	1.21	6	0.90	10	1.02
- Celebrity	3	4.92	0	0.00	6	0.90	9	0.92
Business	0	0.00	13	5.24	25	3.74	38	3.89
Concerns	2	3.28	10	4.03	17	2.54	29	2.97
Peripherals/accessories	2	3.28	5	2.02	20	2.99	27	2.76
XR overview	0	0.00	6	2.42	21	3.14	27	2.76
Future	2	3.28	4	1.61	10	1.50	16	1.64
Figurehead	2	3.28	7	2.82	4	0.60	13	1.33
Legal disputes	0	0.00	4	1.61	5	0.75	9	0.92
Development	0	0.00	5	2.02	3	0.45	8	0.82
Company	0	0.00	2	0.81	3	0.45	5	0.51
History	0	0.00	2	0.81	2	0.30	4	0.41
Crime	0	0.00	0	0.00	3	0.45	3	0.31
Fiction	0	0.00	0	0.00	3	0.45	3	0.31
Regulation	1	1.64	0	0.00	1	0.15	2	0.20
Other	1	1.64	5	2.02	6	0.90	12	1.23

Appendix 2
Applications

Table A2.1 Number of articles mentioning each application type per news outlet

Application type	Sun		Guardian		MailOnline		Overall	
	No.	Percent	No.	Percent	No.	Percent	No.	Percent
Videogames	18	29.51	135	54.44	309	46.26	462	47.29
Film/TV/video	5	8.20	49	19.76	130	19.46	184	18.83
Social media and communication	4	6.56	50	20.16	111	16.62	165	16.89
Tourism/travel	6	9.84	35	14.11	92	13.77	133	13.61
Health	3	4.92	45	18.15	71	10.63	119	12.18
Education	2	3.28	43	17.34	62	9.28	107	10.95
Sport	1	1.64	30	12.10	60	8.98	91	9.31
Pornography, teledildonics and sex	14	22.95	19	7.66	42	6.29	75	7.68
Theme park and rides	8	13.11	9	3.63	52	7.78	69	7.06
Art/design	0	0.00	15	6.05	53	7.93	68	6.96
Training	6	9.84	16	6.45	41	6.14	63	6.45
Music	1	1.64	26	10.48	31	4.64	58	5.94
Social change and awareness	0	0.00	26	10.48	32	4.79	58	5.94
Marketing and advertising	1	1.64	24	9.68	30	4.49	55	5.63
Photography/video recording	1	1.64	9	3.63	44	6.59	54	5.53
Retail	1	1.64	14	5.65	36	5.39	51	5.22
Simulation	1	1.64	4	1.61	44	6.59	49	5.02
Journalism	0	0.00	28	11.29	12	1.80	40	4.09
Museum/exhibition/ archive viewing	1	1.64	14	5.65	17	2.54	32	3.28
Space and science	1	1.64	4	1.61	27	4.04	32	3.28
Industrial and workplace management	0	0.00	9	3.63	22	3.29	31	3.17
Documentary	0	0.00	20	8.06	8	1.20	28	2.87
Military and defence	2	3.28	5	2.02	20	2.99	27	2.76
Other	0	0.00	5	2.02	22	3.29	27	2.76

(*Continued*)

Table A2.1 (Continued)

Application type	Sun		Guardian		MailOnline		Overall	
	No.	Percent	No.	Percent	No.	Percent	No.	Percent
Accessibility	0	0.00	4	1.61	18	2.69	22	2.25
Research	2	3.28	7	2.82	13	1.95	22	2.25
Architecture/planning	0	0.00	11	4.44	10	1.50	21	2.15
Web browsing	1	1.64	6	2.42	13	1.95	20	2.05
Real estate	0	0.00	6	2.42	9	1.35	15	1.54
Product development and testing	1	1.64	3	1.21	10	1.50	14	1.43
Organisation	0	0.00	4	1.61	10	1.50	14	1.43
Fitness	0	0.00	2	0.81	12	1.80	14	1.43
Drones	0	0.00	1	0.40	11	1.65	12	1.23
Wellness	0	0.00	5	2.02	6	0.90	11	1.13
Food and drink	0	0.00	2	0.81	9	1.35	11	1.13
Children's toys/interactive stories	0	0.00	6	2.42	5	0.75	11	1.13
Theatre	0	0.00	5	2.02	5	0.75	10	1.02
Automotive support	0	0.00	0	0.00	9	1.35	9	0.92
Crime prevention and justice	1	1.64	2	0.81	5	0.75	8	0.82
Cosmetics	1	1.64	1	0.40	4	0.60	6	0.61
Emergency services	0	0.00	1	0.40	5	0.75	6	0.61

Appendix 3

Quotes and Paraphrased Statements

Table A3.1 Number of articles with at least one quote/paraphrased statement of each type per news outlet

Source type	Sun		Guardian		MailOnline		Overall	
	No.	Percent	No.	Percent	No.	Percent	No.	Percent
Application creator	17	27.87	120	48.39	250	37.43	387	39.61
Device creator	10	16.39	80	32.26	294	44.01	384	39.30
Other industry or general specialist	7	11.48	45	18.15	137	20.51	189	19.34
Other news source	6	9.84	23	9.27	135	20.21	164	16.79
Researcher/analyst	4	6.56	29	11.69	95	14.22	128	13.10
User (general)	11	18.03	14	5.65	93	13.92	118	12.08
Technology industry specialist	5	8.20	15	6.05	57	8.53	77	7.88
General public	7	11.48	17	6.85	49	7.34	73	7.47
XR facilitator	7	11.48	21	8.47	40	5.99	68	6.96
Game industry specialist	1	1.64	11	4.44	43	6.44	55	5.63
Official reports/ documentation	1	1.64	6	2.42	45	6.74	52	5.32
Platform creator	0	0.00	5	2.02	46	6.89	51	5.22
XR industry specialist	1	1.64	16	6.45	23	3.44	40	4.09
Peripheral creator	2	3.28	5	2.02	23	3.44	30	3.07
User (professional)	0	0.00	4	1.61	22	3.29	26	2.66
External journalist/ blogger	1	1.64	7	2.82	18	2.69	26	2.66
Investor/funder	0	0.00	5	2.02	9	1.35	14	1.43
Marketing materials	2	3.28	2	0.81	10	1.50	14	1.43
XR event organiser	1	1.64	5	2.02	8	1.20	14	1.43
Celebrity	3	4.92	5	2.02	5	0.75	13	1.33
XR job advert	0	0.00	0	0.00	10	1.50	10	1.02
Politician	0	0.00	3	1.21	6	0.90	9	0.92
Retailer	2	3.28	0	0.00	6	0.90	8	0.82
Fiction creator	0	0.00	1	0.40	2	0.30	3	0.31
Other article by same publisher	1	1.64	1	0.40	0	0.00	2	0.20
Unclear	8	13.11	23	9.27	70	10.48	101	10.34
Not specified	4	6.56	23	9.27	92	13.77	119	12.18
Other	5	8.20	12	4.84	21	3.14	38	3.89

Appendix 4

Multimedia Attribution

Table A4.1 Number of attributions to each type per news outlet

Attribution type	Sun		Guardian		MailOnline		Overall	
	No.	Percent	No.	Percent	No.	Percent	No.	Percent
Device creator	21	7.12	39	6.82	729	19.31	789	17.00
Agency	41	13.90	107	18.71	624	16.53	772	16.63
Application creator	36	12.20	131	22.90	489	12.95	656	14.13
Stock image	21	7.12	59	10.31	213	5.64	293	6.31
Other news outlet	18	6.10	11	1.92	130	3.44	159	3.43
Other industry specialist	1	0.34	13	2.27	140	3.71	154	3.32
Publisher	6	2.03	24	4.20	72	1.91	102	2.20
General public	2	0.68	8	1.40	67	1.77	77	1.66
Journalist	0	0.00	40	6.99	34	0.90	74	1.59
Social media	12	4.07	1	0.17	54	1.43	67	1.44
General media	17	5.76	3	0.52	38	1.01	58	1.25
Technology industry specialist	2	0.68	1	0.17	52	1.38	55	1.18
XR facilitator	0	0.00	12	2.10	30	0.79	42	0.90
Celebrity	3	1.02	0	0.00	5	0.13	8	0.17
Other	14	4.75	20	3.50	131	3.47	165	3.55
No attribution	89	30.17	47	8.22	862	22.83	998	21.50
Unclear	12	4.07	56	9.79	105	2.78	173	3.73

Appendix 5
Word Usage in All Categories

Table A5.1 Appearance of terms in all categories per news outlet

Category	Sun No.	Sun Percent	Guardian No.	Guardian Percent	Mail No.	Mail Percent	Overall No.	Overall Percent
FRAME-BASED CATEGORIES								
Immersive	48	49.18	399	57.66	1010	56.59	1457	56.40
Advanced and high-quality	30	34.43	101	28.63	369	30.54	500	30.30
Much-anticipated	10	11.48	141	32.66	329	25.90	480	26.71
Successful	11	13.11	97	25.40	122	14.37	230	17.09
Revolutionary and transformative	6	8.20	63	17.74	146	15.42	215	15.56
Affordable	5	6.56	74	18.55	173	15.12	252	15.46
Transcendent	12	11.48	52	14.11	164	16.02	228	15.25
Important	4	6.56	52	15.32	123	12.28	179	12.69
Easy to use	2	3.28	42	11.29	109	12.43	153	11.57
Comfortable	2	1.64	23	5.65	183	14.22	208	11.26
Social	4	3.28	60	13.31	106	8.98	170	9.72
Different and unique	9	14.75	41	13.71	62	6.14	112	8.60
COUNTERFRAME CATEGORIES								
Expensive	2	3.28	26	8.47	80	8.08	108	7.88
Unsuccessful	3	3.28	37	11.29	24	2.99	64	5.12
Trivial	0	0.00	34	10.89	17	1.95	51	4.09
Uncomfortable	2	1.64	13	4.84	15	1.95	30	2.66
Difficult to use	0	0.00	3	1.21	10	1.20	13	1.13
TONE CATEGORIES								
Positive	60	54.10	459	63.31	785	48.80	1304	52.81
Negative	28	18.03	234	44.76	327	23.20	589	28.35
Concerns	38	21.31	220	33.87	251	16.62	509	21.29
Ailments	13	9.84	134	20.16	221	12.13	368	14.02

Appendix 6
Word Usage Over Time

Table A6.1 Percentage of articles per year with at least one word from each category

Category	2012	2013	2014	2015	2016	2017
Advanced and high-quality	33.33	42.42	34.43	30.12	29.08	28.41
Affordable	12.50	6.06	15.57	21.69	17.39	10.23
Ailments	4.17	3.03	17.21	12.65	20.11	7.20
Comfortable	12.50	12.12	14.75	10.24	12.23	8.71
Concerns	8.33	42.42	28.69	19.28	30.98	17.42
Different and unique	8.33	3.03	15.57	11.45	13.86	10.61
Difficult to use	0.00	0.00	0.82	1.81	1.90	2.27
Easy to use	0.00	24.24	7.38	19.28	12.23	7.20
Expensive	0.00	3.03	4.92	6.63	12.77	4.55
Immersive	33.33	24.24	66.39	60.24	58.70	52.27
Important	8.33	15.15	10.66	16.87	13.32	10.23
Much-anticipated	16.67	39.39	28.69	31.33	29.62	18.18
Negative	16.67	30.30	25.41	19.88	28.53	21.97
Positive	41.67	51.52	53.28	51.81	56.79	48.86
Revolutionary and transformative	20.83	12.12	22.13	18.07	16.03	10.98
Social	0.00	3.03	19.67	15.06	11.41	9.09
Successful	12.50	6.06	25.41	23.49	16.85	11.36
Transcendent	16.67	9.09	18.03	16.27	14.40	15.15
Trivial	8.33	6.06	2.46	3.61	5.71	2.27
Uncomfortable	0.00	3.03	3.28	2.41	3.26	2.65
Unsuccessful	0.00	3.03	7.38	7.83	3.53	5.30

Appendix 7

Comparing Word Usage in VR and AR/MR News Articles

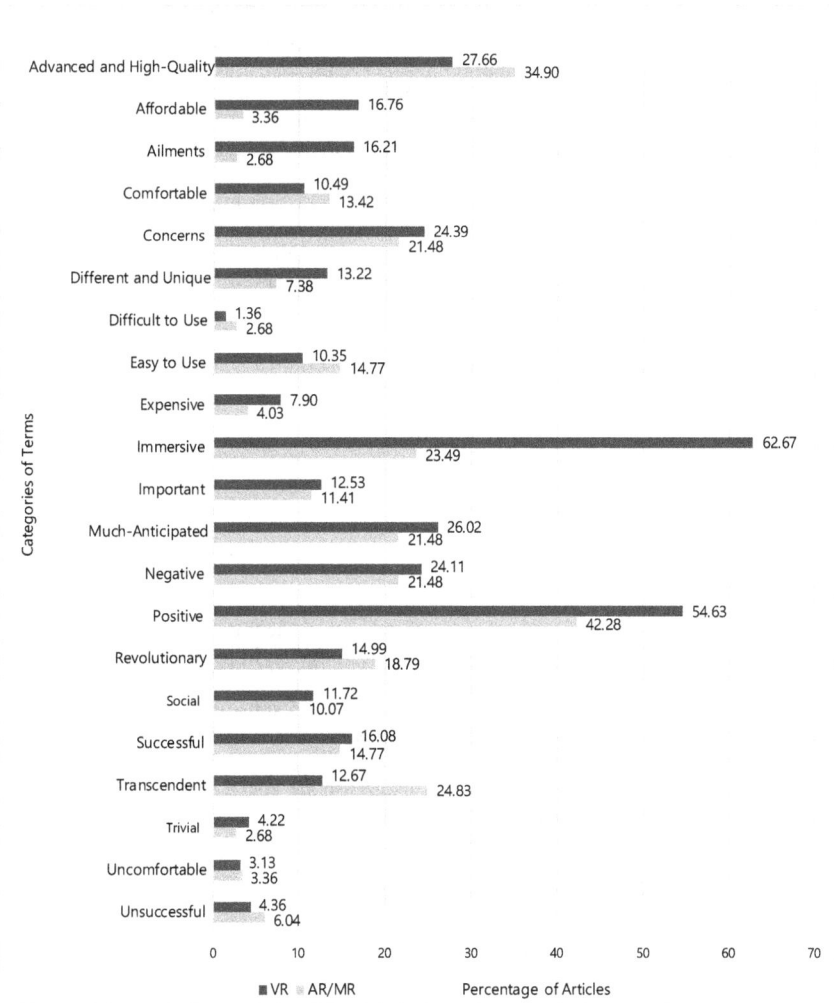

Figure A7.1 Percentage of VR and AR/MR articles with at least one word from each category.

Index

For Product Safety Concerns and Information please contact our EU
representative GPSR@taylorandfrancis.com
Taylor & Francis Verlag GmbH, Kaufingerstraße 24, 80331 München, Germany

www.ingramcontent.com/pod-product-compliance
Lightning Source LLC
Chambersburg PA
CBHW060308220326
41598CB00027B/4268

9 7 8 1 0 3 2 4 5 1 8 8 6